WHO'S WHO IN SKULLS

Papers of the
PEABODY MUSEUM OF
ARCHAEOLOGY AND ETHNOLOGY
HARVARD UNIVERSITY
CAMBRIDGE, MASSACHUSETTS, U.S.A.

VOLUME 82

WHO'S WHO IN SKULLS

Ethnic Identification of Crania from Measurements

William W. Howells

PEABODY MUSEUM OF
ARCHAEOLOGY AND ETHNOLOGY
HARVARD UNIVERSITY
CAMBRIDGE, MASSACHUSETTS, U.S.A.

1995

Cover photographs by Hillel S. Burger

Original design by Janis Owens

Layout by Amy K. Hirschfeld

Production coordination by Margaret R. Courtney

Production assistance by Donna M. Dickerson

Printing and binding by Henry N. Sawyer Company

COPYRIGHT © 1995 BY THE PRESIDENT AND FELLOWS OF HARVARD COLLEGE

ISBN 0-87365-209-6

LIBRARY OF CONGRESS CATALOG CARD NUMBER 95-61421

MANUFACTURED IN THE UNITED STATES OF AMERICA

Contents

FIGURES

TABLES

Acknowledgments

First, to those who made possible the base of information for this and previous studies. To my wife, whose steadfast companionship and patient labor in recording the data allowed the creation of this broad database. Also to those, mostly now living only in memory, whose generosity and hospitality in many places, beginning thirty years ago, opened the opportunity for assembling so much data. I have recorded these debts in detail in earlier publications, and I want now only to say that my gratitude to all these benefactors continues as ever. I recognize more than ever what I owe, especially since some of the work can never be repeated because the material no longer exists. The collection of data was made possible through National Science Foundation grants GS-664 and GS-2465.

This phase of the work has benefited from suggestions and reading by several people, especially G. N. van Vark of Groningen University, Kenneth Jones of Brandeis University, Richard Wright of the University of Sydney, Henry Harpending of Pennsylvania State University, G. Philip Rightmire of Binghamton University, and David Pilbeam and Daniel Lieberman of Harvard University.

In particular, I owe a basic debt to Mary Hyde. Apart from skillful programming she lent her energetic interest to the ideas and purpose of the investigation, was patient with repetitive or varied analyses, sometimes misguided on my part, and was generous with suggestions as to applications supplementing or improving those already tried. With my own limited grasp of such possibilities, I could not have carried out the work without her perceptive help.

1 Introduction

THE PROBLEM

There is a perennial problem of trying to judge the population affinities of a given skull. The skull may be an unidentified modern, but the more interesting case is that of a prehistoric specimen and whether it can be related to peoples of today. One may conclude that the prehistoric specimen is to be affiliated with the present population of its area of origin, but can one make a statement of the probability that this is correct?

An underlying problem: is this last question practicable? I shall be dealing with the craniometric variation within and among populations of *Homo sapiens sapiens*, specifically, living or very recent representatives. The pattern of the outreach of such variation is a matter of interest in itself. Some of my colleagues are explicitly aware of it. It will be present under the surface of much of what follows, and I will address it again in the final discussion.

Regarding the main problem, ethnic identification, the traditional instrument of diagnosis has been the experienced eye. In the present work, it is the use of measurements to express shape. Both methods are "multivariate" as van Vark has several times affirmed (e.g., 1984). Your eye, checking with memory, weights certain visible characters as determining kinship, whereas measurements weight themselves objectively. (As a personal remark, I have never found my own ocular diagnosing particularly satisfying, or as good as that of some colleagues.)

In work like that which follows, the object case—hereinafter referred to as the *target* skull—is placed in an apparent location in a universe of individuals, made up of members of a continuous human species which lacks well-defined subspecies but has clear local tendencies of variation. The detective work involved here is a practical matter, not a theoretical or taxonomic one.

Where the derivation of the target skull is clear or known, such analyses will give mostly satisfactory answers but also, inevitably, a proportion of "wrong" answers. Absolute, positive discrimination is a will-o'-the-wisp, a matter well recognized. Cranially, the populations of the present human species are morphologically too continuous for this end to be possible; they overlap, widely if not seamlessly. This problem is already obvious from attempts to discriminate sex, either visually or by discriminant functions. There is always a substantial zone of overlap. In an exhaustive study using a large series of Portuguese crania of known sex (and the same measurement set as used herein) E. Cunha (1989; see also Cunha and van Vark 1991) found wrong assignments in approximately 14 percent to as much as 20 percent of cases in different attempts. So the question is posed: is a given unknown skull, with a male assignment, indeed a male rather than a somewhat extreme female? There is no way of telling, and that is the problem. It is actually less marked in population discrimination, but it is unavoidable. The problem will be a constant companion in the analysis that follows. (Boulinier, in a little-noticed paper [1970/71] discussed in full the reasons: technical, statistical, and biological.)

PREVIOUS WORK

There has been much use of discriminant functions. Problems like the one stated apparently go back to Rao's (1948) allocation of the Highdown skull as between British Bronze Age and Iron Age samples. Most applications of discriminant functions have been attempts to assign sex, like Cunha's study cited above, and have given results similar to hers. These need not be reviewed. Giles and Elliot (1963) provided one such use of discriminant functions to assign sex (for both American Whites and American Blacks). These authors were also the first to attempt, rather successfully if crudely, cranial discrimination by race (1962), as among "Whites," "Blacks," and "Indians." I followed this lead using Japanese and Ainu crania to allocate Jomon and other prehistoric specimens from Japan (Howells 1966).

In a recent study, Hershkovitz et al. (1990) ran discriminant functions for sex and ethnic group between crania of two Bedouin tribes. Their main purpose was to demonstrate the usefulness of less conventional measurements in such work. In discriminating between the two tribes, they found correct assignment in 95 percent of males and 91 percent of females. A fallacy here lies in judging the efficiency of discrimination from the same samples as those used to derive the functions. In the course of the study herein, when functions were computed as between pooled Europeans and pooled Africans, or other such pairings, correct classification of the same crania was in all cases 100 percent or a point or so less. This is a biased and expected result. The actual problem lies rather in the efficacy of testing specimens not involved in developing the test mechanisms.

Innovations in the use of discriminant functions, complete with programs to run on a personal computer, have recently been developed by others. Professor G. N. van Vark in Holland, who is much at home in mathematics and statistics, has been instrumental with POSCON (created by his colleague D. M. van der Sluis). Richard Wright, Australia's well-known archaeologist, also adept at such techniques, has formulated CRANID (Wright 1992a, 1992b.) Both methods use principal components, rather than discriminant functions, to set up the testing universe for a problem skull. Both use the same database (Howells 1989) as the present effort, which has been carried out in parallel with the others. In fact, the three have been simultaneous endeavors, and I acknowledge the stimulus provided by awareness of their work, though my own work has been independent.

POSCON (used by van Vark) finds Euclidean distances of the target skull from the centroids of the 28 *populations*, testing these with Hotelling's T, and finding P, the posterior probabilities. Wright's CRANID, on the other hand, begins with the principle that the human species is continuous, regardless of tendencies for populations to inhabit more localized regions of the multivariate space. Thus, he sets up a Euclidean space of all the *individuals* of the control populations, placing the target skull in the space, and reading the identity of the 50 nearest individuals as indicating the affiliation of this target skull.

These are highly sophisticated approaches. With them, however, I have not been able, using crania of known identity, to get "correct" answers as often as I would like. Accordingly, what is presented herein has elements in common with van Vark and Wright, but with other steps intended to sharpen the effect. I start from the principle that any stick will do to beat a dog, and therefore that what appears to be effective will be used, whether or not it seems statistically or genetically legitimate.

C. Loring Brace (e.g., Brace and Tracer 1992) has lately been doing similar work but with a different suite of measurements and with ad hoc groupings of crania to represent general or regional populations (e.g., Jomon, Northeast Asia, Northwest Europe, etc.). Using distances based on discriminant analysis, he finds relative probabilities of affiliation for given test skulls, i.e., posterior probabilities as with POSCON. There is an easy misunderstanding here. In such analyses, there are prior and posterior probabilities of assignment, which assume that a target specimen does, or could, belong to one of the populations involved (see Albrecht 1992). Prior probabilities are normally equal for all populations and sum to 1.0; that is, if there are 4 reference populations, the prior probabilities are 0.25 for each. The posterior probabilities, resulting from discriminant scoring, also are normally made to sum to 1.0, assuming as above that any specimen belongs to one of the same populations. (This is how van Vark, below, uses posterior probabilities.) However, the *absolute* probability, as computed herein, is different; a specimen may be so distant, in statistical terms, from all populations that its probability of assignment to the nearest population is 0.01 or less, while its posterior probability may still be 1.00 or close to it, a deceptive reading. Some of Brace's results will be given further on.

MATERIALS

The purpose of my report *Cranial Variation in Man*, published in 1973, was the comparison, by multivariate analyses of crania, of a number of living or quite recent indigenous populations widely varied geographically as well as "racially," i.e., in outward appearance of the living. I decided on a list of 57 measurements and a generous sample size of 50 specimens of each sex for every population included. This meant visiting collections comprising about three times that number in order to find the needed number of specimens in sufficiently complete condition. Beginning in 1965, I devoted a number of months, mostly in travel, to amassing the desired data on a total of 17 populations.

This allowed the relating of the populations by objective statistical methods as well as some conclusions as to the morphology involved. Encouraged by the results, I decided to expand the base in order to achieve more even geographical coverage, particularly so as to represent each of six major world areas by three separate populations. I ended, in fact, with a total of 28 series, in every case definable as a natural grouping and not constituting a conglomeration such as "Japanese" or "American Indian." This allowed, in my 1989 report, *Skull Shapes and the Map*, general statements about some apparent characters distinguishing major areas as well as degrees of variation within such regions.

The total corpus of this carefully selected material includes 1,348 male crania and 1,156 female crania. Necessarily, this universe is not ideal and species-wide but only a partial one, limited by the availability of collected material; it has spacious interstices for which no information is included, notably, all of southern Asia. This becomes especially evident in trying to place skulls whose natural place would be in such a lacuna. There is no cure for this.

Nevertheless, the data body of 2,504 crania is a substantial one in terms of coverage, and it has proven useful to others as well as myself. This data body is the matrix into which we here try to insert a further skull (or rather its shape as conveyed by the 57 measurements taken on it). The procedures I have developed are necessarily multivariate statistical, of two principal kinds. One kind (DISPOP) uses the *centroids of populations* as the pivots, as in POSCON used by van Vark, attempting to locate the target skull with respect to these. The other (POPKIN) uses the *target* as the pivot, and locates other known individuals with respect to this skull (see also Wright 1992a, 1992b.) It goes without saying that the number of possible treatments in detail is infinite. Those used herein have been selected after a good deal of experimentation and inspection, which will be referred to in a few places only.

As a small example, subsets of the measure-ments (e.g., facial versus vault) can be tried and have been tried in the course of this work. In no case have these appeared to be as efficient as the total set, which accordingly is used throughout.

POPULATIONS

The samples, described in detail in Howells (1989), are here given thumbnail characteriza-tions as follows:

Europe

Norse. From a few medieval parish graveyards in the vicinity of Oslo, Norway. University of Oslo.

Zalavár. From cemeteries at this site in western Hungary, dating to the ninth to eleventh cen-turies, comprising a cosmopolitan population of Romanized Germans, Slavs, and Franks. Natural History Museum, Budapest.

Berg. From the charnel house of a small village in Carinthia, Austria, essentially comprising all the individuals of about five generations of the village. This would be the genetically most com-pressed of the population samples. American Museum of Natural History, New York.

Egypt, Gizeh. A natural "Caucasoid" member of this group, as will appear in results. From a single large cemetery south of the pyramids dat-ing from the twenty-sixth to thirtieth dynasties (ca. 600–200 B.C.) excavated for Karl Pearson by Flinders Petrie. Duckworth Laboratory, Cam-bridge University.

Africa

Teita. A Bantu-speaking tribe in southeastern Kenya. Duckworth Laboratory, Cambridge University.

Dogon tribe, Mali, West Africa. Musée de l'Homme, Paris.

Zulu, South Africa. Dissecting room and archaeological specimens identified as Zulu, of known sex. R. A. Dart Collection, University of the Witwatersrand.

Bushman (San), Cape Province, South Africa. A general grouping of identifiable or persuasively probable Bush individuals from various collec-tions, mostly the Anthropological Institute, Vien-na, and the South African Museum, Cape Town.

Australo-Melanesia

South Australia, Lower Murray River. From two closely related tribes in a long-stable area. South Australian Museum, Adelaide.

Tasmania, general. Not a local population but an assemblage of crania from the whole island where documentation as Tasmanian aboriginal seemed acceptable. These remains have since been almost or entirely destroyed by cremation at the demand of part-aboriginal peoples of Tasmania.

Tolai, northern New Britain. A well-defined Austronesian-speaking people of the region south of Rabaul, collected early in this century by Richard Parkinson. This part of the collection was acquired by the American Museum of Natur-al History, New York.

Polynesia

Hawaii, Mokapu Peninsula, Oahu. From bur-ial plots believed to date from the late precontact period, from about A.D. 1400 to 1790. Excavated mostly from 1938 to 1940. Bishop Museum, Honolulu. These remains are in the process of reburial at the demand of native Hawaiians.

Easter Island. Skulls believed to date from about A.D. 1100 to 1868, with no differences dis-cernible between early and later elements. Those used are mostly in the Musée de l'Homme, Paris, and the Natural History Museum, London.

Moriori, Chatham Islands. A long isolated Polynesian population, now extinct. Specimens largely in the Natural History Museum, London, and the University of Otago, New Zealand.

Americas

Arikara. From a village in South Dakota believed to have been occupied by protohistoric Arikara from about A.D. 1600 to 1750. U. S. National Museum and University of Kansas.

Santa Cruz Island, California. Apparently Chumash tribe, of recent centuries, and thus a fairly isolated population. Peabody Museum, Harvard University, and Natural History Museum, Washington, D.C.

Peru, Yauyos district, southeast of Lima. Collected during archaeological work by J. C. Tello and apparently all precontact. Peabody Museum, Harvard University.

Far East

North Japan, Hokkaido. From the dissecting room collection of the Department of Anatomy, Hokkaido University Medical School. Of known age and sex.

South Japan, Kyushu. From five prefectures of northern Kyushu. From the dissecting room collection of the Kyushu University Department of Anatomy. Of known age and sex.

Hainan Island, China. From burials of the Chinese immigrant population, essentially from South China, excavated by T. Kanaseki. Academia Sinica and National Taiwan University, Taiwan.

Anyang, Bronze Age Chinese (no female sample). From sacrificial burial pits of Shang Dynasty tombs; actual makeup of the population is not clear but perhaps moderately cosmopolitan. Academia Sinica, Nankang, Taiwan.

Other

Philippine Islands (no female sample). A general series composed of convicts who died in prison and were not claimed for burial. Probably representative of the Philippines generally. Medical School, University of the Philippines, Manila.

Atayal, aboriginal tribe of Taiwan. Probably a generally isolated population genetically. Academia Sinica and National Taiwan University, Taiwan.

Guam, Latte period. Excavated from around various latte sites, mainly around Tumon Bay. The date is about A.D. 1000 and definitely pre-Spanish. Bishop Museum, Honolulu.

Ainu, Hokkaido. Largely from abandoned cemeteries in good Ainu territory in southern and southeastern Hokkaido, often certified as to ethnicity and sex by grave goods. Faculty of Medicine, University of Hokkaido and University Museum, Tokyo.

Andaman Islands. A small population, geographically isolated. From many collections, principally the Natural History Museum, London, and Cambridge University.

Eskimo, Greenland. From West and Southeast Greenland, burials associated with the Inugsuk culture, before Danish colonization (ca. A.D. 1750) but following early Norse settlements; graves show no evidence of European contact. Anthropological Laboratory, University of Copenhagen.

Buriats, Siberia. From the region around the southern end of Lake Baikal. Institute of Ethnography, St. Petersburg.

The male samples of the above populations number from 29 to 58 individuals (see table 1), mostly close to 50. The females samples number from 18 to 55, though again mostly close to 50.

TEST specimens

In addition, the material includes over 600 TEST specimens, preponderantly male. These are the target skulls, or object specimens, of the present investigation. This set includes some extra individuals from the same populations as the base populations. In addition, and more important, are: single examples or small series from a variety of other populations; individuals of doubtful affiliation; and prehistoric skulls. All specimens were sufficiently complete to allow the full suite of measurements to be made or fairly closely estimated.

APPROACHES TO THE DATA

As I have said above, two kinds of analysis can be used. The first, which I call DISPOP, attempts to estimate which population centroid is closest to, or least removed from, the position of the target specimen. The second, which I call POPKIN, places the individuals, of whatever population, in order of their distance from the object. This is more elaborate, but gives similar answers. Because of some limitations in the female data I have applied POPKIN only to the males.

In my 1973 work, done more than 20 years ago, the data were subjected to a single factor analysis. Today, with the ridiculous speed of present computers, it would be irresponsible not to investigate numerous treatments and analyses, with the problem being, not getting enough solutions, but rather choosing among them effectively. Many examinations have been performed which do not appear here at all.

Two basic treatments have been applied throughout, in preparation for both DISPOP and POPKIN.

C-scores

C-scores are intended to control the factor of size. They are derived as described in *Skull Shapes* (1989) by a type of double-centering. For the first six main geographic areas listed above, the constituent 18 samples (excluding Egypt, Bushman, and Anyang) are treated as the *basic* or *core* series for worldwide coverage, leaving aside such special populations as Andaman Islanders, Ainus, or Eskimos. From these were obtained *a)* the mean of the 18 population means (not the grand mean, to avoid weighting by different sample numbers) and *b)* a standard deviation based on the pooled variances of the same populations. These base figures were used to render all individual raw scores into Z-scores, or standard scores; i.e., the deviation of each score from the general mean is divided by the general standard deviation, so that over all spec-

imens the mean approximates zero and the standard deviation approximates 1.0.

(The results are "approximate" because the actual grand mean was not used, and because specimens outside the core series are not involved. Note that here, and in all subsequent data, the standard deviation is actually not 1.0 but 10.0. All Z-scores were arbitrarily multiplied by 10.0 at the beginning, for readability, in order to avoid having too many of the significant digits to the right of the decimal point. See Howells 1973, 1989.)

Then the mean was found for the 57 Z-scores for each single individual, taken as a measure of its relative overall size. This figure (which I have called PENSIZE) was then subtracted from all of that individual's Z-scores, giving it a personal mean, for its C-scores, of zero. With size thus supposedly made constant, these scores ideally reflect shape only: a long-skulled individual will be positive in the C-score for GOL, length, and negative in XCB, breadth, regardless of absolute size. (See also Howells 1986, 1989.) All these operations, of course, were carried out separately for males and females.

This kind of size control is subject to a couple of queries. Size in itself may be a character of significance, but is here supposedly eliminated. Also, as will appear later, size is apt to be associated with robusticity. In a number of significant prehistoric specimens which are large and robust, size is removed but the robusticity remains, with what seem to be significant effects in the analysis. (See Addendum.)

Canonical Variates

A further general treatment is the scoring of the data on canonical variates.[1] It was decided, after some experiments and examinations, that this would, as should be expected, segregate the

---

1. The procedure used was CANDISC, of the SAS package.

populations best and provide the most explicit space for the placing of target skulls. This element of discrimination has been avoided by both van Vark and Wright, who use principal components instead.

Accordingly, in the bulk of what follows, distances and other comparisons are based on canonical variates computed on the 28 male or 26 female samples, at whatever stage is appropriate. Specifically, the first 14 such discriminants are used throughout. This choice results from inspection of trials using other numbers of discriminants, from 6 to 22 (out of a possible maximum of 27), but in addition has a rationale, also experimental. A number as low as 6 discriminants makes a good division among major geographical groupings—this corresponds with what was found in the analyses in *Skull Shapes* (1989). More subtle distinctions or affiliations among populations, however, seem to escape this limited number of variates. Later variates, i.e., above 6 and approaching 14, are extremely difficult to interpret morphologically or in terms of population groups. Nevertheless, as demonstrated by

close agreement between the separate analyses of male and female series in *Skull Shapes*—an important point[2]—it is apparent that such later order variates are not simply noise and eraser dust but in fact carry real, if obscure, information. (See Albrecht 1992.)

(It is always possible to rotate such axes to positions in which each is more specific and readable for morphological reference. This has not been done here for two reasons: the functions become overly specific, and in fact often express one measurement or a very few, conveying little about more general morphology; and in hierarchically ranked, unrotated functions the bulk of the information is automatically found in those extracted first, and even if some are unreadable, such functions are the most effective in finding distances.)

In general, then, each individual is represented by a vector of 14 canonical variate scores. Each population has a centroid, a vector of 14 means of the scores of its constituent individuals. Euclidean distances were computed among these ad hoc.

---

2. Wolpoff (1994) has made the mystifying assertion that my measurement set, and multivariate analyses, are exposed as inadequate, in part because "the patterns of relationships for males and females invariably differ, often dramatically." Quite the contrary; the most casual inspection of the figures and tables in *Skull Shapes* will demonstrate the close correspondence of the sexes in several independent analyses, and I remarked then that this close agreement was important in demonstrating the general reliability of the results.

2 DISPOP

Table 1 shows, for each population separately, the mean of the *distances* of the individuals from the centroid of that population, and also the standard deviation of those distances. These values differ somewhat among populations, but seemingly not systematically, as inspection of the two sexes suggests. In some cases the female mean, or the standard deviation, is higher than the male; in other cases the opposite. Also, the values do not seem to reflect any special effects among populations of supposed high genetic isolation, such as Berg, Moriori, or Santa Cruz.

There is one rather puzzling exception, the Tasmanians, who show distinctly higher mean differences and variation in both sexes. This phenomenon was not obvious to me from figures in *Skull Shapes*, but is probably real. If there exists an overall greater variation in the sample, this might be due to the breadth of the area of selection, combined with errors of inclusion (discussed in *Skull Shapes*). However, such excess variation is not seen in some other samples assembled over a broad territory, like Bushmen or Filipinos. In an illuminating article on Tasmanian crania, Pardoe (1991) finds that, in spite of 8,000 years of genetic isolation of that population from the Australian continent, Tasmanian skulls deviate, in non-metric trait distances, no more from mainland series than do such local groupings from one another. Whatever the reason may be, it is not likely to be discovered. All surviving Tasmanian osteological materials have been removed from museums and turned over, by law, to the island's living part-aboriginals for disposal.

Computationally, taking account of the differing degrees of variation in estimating the probable significance of differences would not be a problem. But, given the nature of the material itself, and the context of the treatment, it is easy to be excessively precise. Thus it seems wisest to disregard these apparent fluctuations in variation, and to bring all distances to a common footing, as below. This applies to the Tasmanians similarly because, if computation took account of their higher standard deviations, this would result in a more inclusive reach over the distances of other skulls from the Tasmanian mean, and thus an apparently higher probability of identification as "Tasmanian" than for other populations. This would hardly be desirable or logical.

Estimation of Probability

A generalized within-group standard deviation for the distances was computed by summing the sums of squared deviations within each group, in a manner comparable to that used originally to compute Z-scores (Howells 1989). These generalized standard deviations are 0.738 for the males and 0.697 for the females. (The mean of

TABLE I

Distribution of individual distance from population centroids

Group	Mean	Standard Deviation	N	Mean	Standard Deviation	N
	MALE			FEMALE		
Norse	3.59	0.71	55	3.49	0.70	55
Zalavar	3.50	0.72	53	3.64	0.76	45
Berg	3.76	0.74	56	3.94	0.73	53
Teita	3.87	0.74	33	3.82	0.68	50
Dogon	3.88	0.64	47	3.78	0.63	52
Zulu	3.70	0.70	55	3.83	0.62	46
Australia	3.50	0.71	52	3.45	0.50	49
Tasmania	4.13	1.13	45	4.07	1.13	42
Tolai	3.63	0.79	56	3.24	0.66	54
Hawaii	3.65	0.70	51	3.50	0.79	49
Easter I.	3.59	0.57	49	3.56	0.71	37
Moriori	3.74	0.78	57	3.60	0.70	51
Arikara	3.44	0.76	42	3.65	0.80	27
Santa Cruz	3.39	0.77	51	3.30	0.64	51
Peru	3.30	0.70	55	3.45	0.65	55
N Japan	3.80	0.86	55	3.81	0.63	32
S Japan	3.53	0.67	50	3.42	0.71	41
Hainan	3.51	0.76	45	3.85	0.62	38
Atayal	3.60	0.68	29	3.61	0.59	18
Philippines	3.55	0.69	50			
Guam	3.30	0.70	30	3.40	0.66	27
Egypt	3.45	0.59	58	3.38	0.65	53
Bushman	3.75	0.87	41	3.80	0.70	49
Andaman Is	3.77	0.83	35	3.60	0.84	35
Ainu	3.66	0.72	48	3.77	0.58	38
Buriat	3.95	0.79	55	3.77	0.76	54
Eskimo	3.72	0.62	53	3.68	0.70	55
Anyang	3.46	0.79	42			

the population mean distances—those in table 1—is 3.62 for males and 3.63 for females.)

Figure 1 shows a plot of the 1,348 male individuals, as distances from their own group means in terms of the generalized male standard deviation of 0.738. The individual distances are taken from their appropriate group means. The resulting deviations are divided by the general standard deviation of 0.738 to form a curve of distribution around the group mean of zero. These distributions have been cumulated by being superimposed to show a total pooled sample (with a mean of zero). The resulting distribution is graphed for the total of 1,348 individuals, using a class interval of 0.1, now having a mean of zero and a (theoretical) standard deviation of 1.0. This is plotted against a normal curve constructed for the same number, mean, standard deviation, and class interval. The fit appears good enough to justify the use of the computed

standard deviations to estimate probabilities in the case of test skulls. There appear to be minor degrees of kurtosis and skewness in the distribution of distances, but hardly enough to take into account.

Accordingly, DISPOP proceeds as follows: a) find the Euclidean distance, D, of the individual target skull from the centroid of each population; b) test this against the generalized standard deviation, viz., $T = (D - M)/SD$; c) from T, read the probability, P, from the normal curve, that the target might be a member of the population in question. (Note that this is not a "posterior probability," in which the probabilities across all the populations sum to 1.0, but is the absolute probability based on the distribution of distances within the population. Here, the absolute probabilities across the populations do not sum to 1.0. Generally the sum is less but often more.)

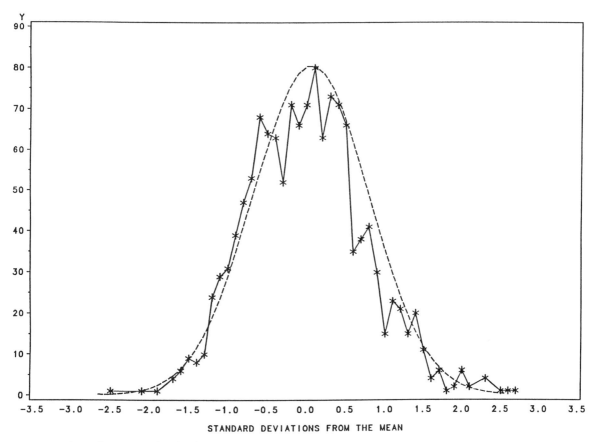

Figure 1. Plot of 1,348 individuals against theoretical curve.

In the tables following, the distance, D, and the probability, P, are shown on successive lines and, for a given skull, the populations are listed in order of distance. Only the closest populations are shown (10 in tables 2 and 3, 8 in tables 4 and 5), since P drops off quickly to vanishingly low values—in the tables, any P value less than .0005 is shown as a dash, not as .00. I regret that the numerals in these tables are so small, and smaller still under POPKIN and later tables. It is preferrable to have the table upright and complete on most pages, at the price of a little eyestrain, since the figures do not call for very exact study.

Peru tests

To test the efficacy of DISPOP in locating a skull, we may use a Peruvian sample as a particu-

lar data set. From the whole Peruvian collection, which is an ample one, a surplus of specimens was originally measured, with a view toward just this kind of testing. As related in *Cranial Variation in Man* (1973), 75 skulls of each sex were included, with 55 being retained as the Peruvian reference sample. The other 20 were placed in the TEST series. These were largely picked out at random, as every third or fourth in the numerical order of ID numbers. However, this is not a completely random set; a very few were put in the test series because I thought they were unusual in some respect.

These TEST specimens were, of course, excluded from all computation of discriminants, interpopulation distances, etc. Table 2 shows the DISPOP results for the 40 test skulls, results which are highly satisfactory in general while substantiating the warning that occasional cases may be

TABLE 2

DISPOP results: Peru TEST cases: males

1403 Peru

Morior	Arikar	Tolai	S.Cruz	Peru	Tasman	N.Jap.	Hainan	Anyang	Eskimo
6.5	6.7	6.8	6.8	7.4	7.5	7.7	8.1	8.2	8.2
-	-	-	-	-	-	-	-	-	-

1410 Peru

Peru	Guam	Zalvar	Norse	S.Cruz	Hainan	Phil	Anyang	Atayal	Egypt
3.3	4.7	4.7	4.7	4.8	4.8	5.0	5.1	5.2	5.2
0.65	0.08	0.07	0.06	0.05	0.05	0.03	0.03	0.02	0.01

1414 Peru

Peru	S.Cruz	Arikar	Hainan	Norse	Zalvar	Phil	Atayal	Berg	Guam
1.8	4.1	4.2	4.9	4.9	5.0	5.2	5.3	5.4	5.6
0.99	0.24	0.20	0.04	0.04	0.03	0.02	0.01	0.01	0.00

1418 Peru

Peru	S.Cruz	Andamn	Egypt	Phil	Zalvar	Norse	Hainan	Atayal	Teita
3.8	4.3	4.6	4.7	4.7	4.9	4.9	5.1	5.2	5.2
0.41	0.20	0.09	0.07	0.07	0.04	0.04	0.02	0.02	0.02

1422 Peru

Peru	Arikar	Zalvar	Hainan	Atayal	Berg	Phil	S.Cruz	Guam	Norse
3.1	4.5	4.7	4.8	4.9	4.9	5.0	5.2	5.3	5.7
0.77	0.11	0.07	0.05	0.04	0.04	0.03	0.02	0.01	0.00

1426 Peru

Peru	Zalvar	S.Cruz	Atayal	Hainan	Phil	Norse	Guam	Berg	S.Jap.
3.6	4.9	5.3	5.5	5.6	5.7	5.7	5.7	5.8	6.0
0.50	0.04	0.01	0.01	0.00	0.00	0.00	0.00	0.00	0.00

1430 Peru

Peru	S.Cruz	Morior	Arikar	Hawaii	Norse	N.Jap.	Zalvar	S.Jap.	Guam
3.6	4.8	5.4	5.4	5.7	5.8	5.9	6.0	6.1	6.1
0.49	0.05	0.01	0.01	0.00	0.00	0.00	0.00	0.00	0.00

1435 Peru

Berg	Zalvar	Norse	Egypt	Andamn	Peru	Atayal	S.Cruz	Phil	Arikar
3.3	3.6	4.3	4.8	4.9	5.3	5.3	5.4	5.5	5.5
0.68	0.51	0.17	0.06	0.04	0.01	0.01	0.01	0.01	0.00

1439 Peru

Peru	S.Cruz	Zalvar	Norse	Arikar	Guam	Hawaii	Berg	Tolai	N.Jap.
3.7	5.1	5.9	6.1	6.5	6.5	6.5	6.6	6.6	6.7
0.47	0.02	0.00	0.00						

1442 Peru

S.Cruz	Arikar	Peru	Norse	Zalvar	Guam	Morior	Berg	Phil	N.Jap.
4.2	4.2	4.3	6.3	6.3	6.4	6.4	6.6	6.6	6.8
0.23	0.21	0.18	0.00	0.00	-	-	-	-	-

1446 Peru

Peru	S.Cruz	Arikar	Zalvar	Hainan	Norse	Berg	Phil	Atayal	S.Jap.
4.1	5.4	5.5	6.0	6.1	6.3	6.3	6.6	6.7	6.8
0.27	0.01	0.01	0.00	0.00	0.00	0.00	-	-	-

1450 Peru

Peru	Arikar	Norse	Zalvar	Berg	S.Cruz	Atayal	Hainan	S.Jap.	Anyang
3.5	4.3	4.4	4.6	5.1	5.1	5.3	5.4	5.6	5.7
0.55	0.16	0.13	0.10	0.02	0.02	0.01	0.01	0.00	0.00

1749 Peru

Peru	Hainan	S.Cruz	Arikar	Phil	N.Jap.	S.Jap.	Anyang	Atayal	Andamn
4.4	5.7	5.8	6.0	6.3	6.4	6.5	6.6	6.6	6.6
0.15	0.00	0.00	0.00	0.00	-	-	-	-	-

1750 Peru

Peru	Atayal	S.Cruz	Arikar	Hainan	Andamn	Tolai	Phil	Norse	Zalvar
3.3	4.5	4.7	4.8	5.0	5.3	5.5	5.6	5.7	5.7
0.64	0.11	0.07	0.05	0.03	0.01	0.01	0.00	0.00	0.00

1751 Peru

Peru	S.Cruz	Arikar	Tolai	Phil	Hainan	Guam	Atayal	Tasman	Zalvar
3.1	4.7	4.9	5.3	5.4	5.6	5.8	5.9	5.9	6.0
0.75	0.07	0.04	0.01	0.01	0.00	0.00	0.00	0.00	0.00

1752 Peru

Peru	Norse	Zalvar	S.Cruz	Atayal	Hainan	Berg	Phil	Egypt	S.Jap.
3.4	4.6	4.6	4.9	4.9	5.0	5.1	5.4	5.4	5.7
0.63	0.10	0.09	0.04	0.04	0.03	0.03	0.01	0.01	0.00

1753 Peru

Peru	Hainan	Anyang	Norse	S.Jap.	Phil	Morior	N.Jap.	Arikar	Atayal
4.0	4.6	4.7	4.9	5.2	5.2	5.3	5.4	5.4	5.4
0.29	0.10	0.08	0.04	0.02	0.01	0.01	0.01	0.01	0.01

1754 Peru

Peru	S.Cruz	Berg	Zalvar	Arikar	Norse	Phil	Atayal	Hainan	Egypt
4.3	5.3	5.3	5.6	6.2	6.3	7.1	7.2	7.3	7.4
0.18	0.01	0.01	0.00	0.00	0.00	-	-	-	-

1755 Peru

Andamn	Peru	Atayal	Hainan	Egypt	Phil	Anyang	Norse	S.Jap.	S.Cruz
3.9	4.3	4.7	4.9	4.9	5.2	5.3	5.4	5.4	5.4
0.35	0.18	0.07	0.05	0.04	0.02	0.01	0.01	0.01	0.01

1756 Peru

Peru	Zalvar	Berg	Norse	Phil	Arikar	Atayal	Hainan	S.Cruz	Guam
4.6	5.0	5.0	5.4	5.6	5.8	6.0	6.1	6.1	6.2
0.08	0.03	0.03	0.01	0.00	0.00	0.00	0.00	0.00	0.00

TABLE 2

DISPOP results: Peru TEST cases: females

1459 Peru

Peru	Arikar	Berg	Norse	Zalvar	Zulu	S.Cruz	Egypt	Teita	Atayal
4.3	5.6	5.7	6.0	6.0	6.3	6.4	6.5	6.6	6.8
0.16	0.00	0.00	0.00	0.00	-	-	-	-	-

1461 Peru

Peru	S.Cruz	Atayal	Arikar	Zalvar	Berg	S.Jap.	Hainan	Norse	Andamn
2.7	4.3	4.6	4.6	4.8	5.0	5.3	5.3	5.4	5.7
0.90	0.15	0.09	0.08	0.05	0.03	0.01	0.01	0.00	0.00

1465 Peru

Peru	Arikar	Berg	Zalvar	S.Jap.	S.Cruz	Atayal	N.Jap.	Norse	Hainan
2.5	4.2	5.0	5.1	5.5	5.6	5.7	5.7	5.8	6.0
0.95	0.22	0.03	0.02	0.00	0.00	0.00	0.00	0.00	0.00

1469 Peru

Arikar	Peru	Norse	Berg	Zalvar	Egypt	Zulu	S.Cruz	Ainu	Teita
4.6	4.7	4.7	5.4	5.4	5.8	6.0	6.1	6.6	6.6
0.07	0.07	0.07	0.01	0.01	0.00	0.00	0.00	-	-

1473 Peru

Peru	Andamn	Egypt	Arikar	Zalvar	Dogon	Norse	Berg	S.Cruz	Ainu
4.9	5.5	5.8	5.8	6.0	6.1	6.2	6.2	6.3	6.4
0.03	0.00	0.00	0.00	0.00	0.00	0.00	-	-	-

1477 Peru

Peru	S.Jap.	Arikar	N.Jap.	Hawaii	Hainan	Atayal	Zalvar	Tolai	Ainu
4.1	4.5	4.9	4.9	5.1	5.1	5.3	5.6	5.6	5.7
0.23	0.10	0.04	0.03	0.02	0.02	0.01	0.00	0.00	0.00

1481 Peru

Peru	Arikar	Atayal	S.Cruz	Hainan	S.Jap.	Guam	N.Jap.	Andamn	Berg
4.1	4.3	4.6	4.9	5.1	5.2	5.4	5.7	5.7	5.9
0.23	0.18	0.09	0.04	0.02	0.01	0.00	0.00	0.00	0.00

1485 Peru

Peru	Arikar	Norse	Egypt	Hainan	Zalvar	S.Jap.	Guam	Ainu	N.Jap.
3.5	4.3	4.5	4.7	4.8	4.8	4.9	5.1	5.1	5.4
0.57	0.19	0.10	0.06	0.05	0.05	0.04	0.02	0.02	0.00

1489 Peru

Peru	Arikar	Andamn	Hainan	Atayal	S.Cruz	Berg	Zalvar	Egypt	S.Jap.
3.2	4.6	5.0	5.1	5.2	5.3	5.4	5.7	5.8	5.8
0.74	0.07	0.03	0.01	0.01	0.01	0.01	0.00	0.00	0.00

1493 Peru

Peru	Hainan	Arikar	S.Jap.	Guam	S.Cruz	Egypt	Norse	N.Jap.	Zulu
5.0	6.2	6.3	6.4	6.7	6.7	6.8	6.9	7.0	7.1
0.03	0.00	-	-	-	-	-	-	-	-

1497 Peru

Peru	Arikar	Zalvar	Norse	Berg	S.Jap.	S.Cruz	Tolai	Atayal	Ainu
3.5	4.5	5.2	5.5	5.7	5.7	5.8	5.9	5.9	6.1
0.59	0.11	0.01	0.00	0.00	0.00	0.00	0.00	0.00	0.00

1501 Peru

Arikar	Peru	Berg	S.Cruz	Guam	Ainu	Hainan	Morior	S.Jap.	Atayal
5.0	5.1	6.6	6.7	6.8	6.8	6.9	6.9	7.0	7.0
0.02	0.02	-	-	-	-	-	-	-	-

1505 Peru

Peru	S.Cruz	Arikar	S.Jap.	Hainan	Atayal	Berg	Zalvar	N.Jap.	Norse
2.8	5.0	5.2	6.3	6.5	6.6	6.6	6.7	6.7	6.9
0.89	0.03	0.01	-	-	-	-	-	-	-

1757 Peru

Peru	Hainan	Arikar	S.Jap.	Atayal	S.Cruz	Guam	N.Jap.	Zalvar	Berg
2.7	4.2	4.3	4.4	4.5	4.8	5.0	5.2	5.4	5.7
0.92	0.20	0.16	0.13	0.11	0.05	0.03	0.01	0.01	0.00

1758 Peru

Peru	Arikar	S.Cruz	S.Jap.	Hainan	Atayal	N.Jap.	Morior	Guam	Berg
4.2	4.9	5.6	6.7	6.8	6.8	6.9	7.0	7.0	7.1
0.20	0.04	0.00	-	-	-	-	-	-	-

1759 Peru

Peru	Arikar	Zalvar	Berg	Norse	S.Cruz	Egypt	S.Jap.	Atayal	N.Jap.
3.2	4.6	4.7	5.0	5.0	5.1	5.6	5.6	5.8	5.8
0.75	0.09	0.06	0.03	0.03	0.01	0.00	0.00	0.00	0.00

1760 Peru

Peru	S.Cruz	Arikar	Berg	Zalvar	S.Jap.	Atayal	Norse	N.Jap.	Hainan
4.4	5.9	5.9	6.5	6.6	6.9	7.0	7.3	7.3	7.5
0.15	0.00	0.00	-	-	-	-	-	-	-

1761 Peru

Arikar	Peru	S.Cruz	Berg	Hainan	S.Jap.	Zalvar	N.Jap.	Atayal	Norse
3.6	3.6	4.5	4.8	4.8	5.4	5.4	5.5	5.5	5.5
0.51	0.49	0.10	0.05	0.04	0.00	0.00	0.00	0.00	0.00

1762 Peru

Peru	Arikar	S.Cruz	Berg	Buriat	Zalvar	Hainan	Norse	Guam	Atayal
4.8	5.9	6.7	7.0	7.1	7.4	7.8	7.8	8.0	8.1
0.04	0.00	-	-	-	-	-	-	-	-

1763 Peru

Arikar	Peru	S.Jap.	Hainan	Guam	Atayal	Ainu	Hawaii	Morior	Zalvar
3.7	3.8	4.4	4.7	4.8	4.9	4.9	4.9	5.0	5.2
0.47	0.43	0.13	0.07	0.05	0.04	0.03	0.03	0.03	0.01

quite off the mark. Among males, the first specimen (*#1403*) on the list appears as closest to
Moriori, but at a distance making it improbable
that it should be assigned to that or any population. However, this skull has an abnormal development of attachment for the temporal muscles,
giving a high maximum frontal breadth, and a
shape which is thus exceptional.

Number *1435* is firmly European in diagnosis,
with a probability of approximately 1 percent of
being Peruvian (see further discussion of this case
under POPKIN). Number *1755* is closest to
Andamanese, with an 18 percent chance of
assignment to Peru. In a blind test, however, it
would appear Andamanese if this did not seem to
be geographically absurd. All the remaining 17
skulls are read as "Peru" except for *#1442*,
which at least is clearly American Indian with
Peru at an 18 percent probability.

Among the 20 females, results are even more
satisfactory. All are "Indian": 16 are diagnosed
as Peru, 4 as Arikara. The last three examples in
the list were originally excluded from the main
series, not at random like the rest, but because of
slight oddities of shape. Two of the "Arikara"
are placed here. In these cases, the probability for
"Peru" is also high. And, for what it is worth,
Arikara is second choice in 9 out of 20 cases.

We may compare the figures with results given
by POSCON, kindly furnished by Professor van
Vark. In this case, 14 principal components were
computed on 57 measurements and the analysis
has been applied to the 20 male Peru TEST
skulls. The readings give 11 Peru and 2 Arikara,
for 13 "Indians," but also 3 Andamanese, 2
Norse, 1 Filipino, and 1 Hawaii. "Errors" tend to
involve the same cases as mine (the cases of
#1403, for good reason, *#1435*, and *#1755*), if
not always leading to the same diagnoses. While
these results seem somewhat less good than those
of DISPOP, the difference is not great, and as will
be seen immediately below, results with tested
Maoris are much the same for the two systems.

Maori tests

The Peru test skulls were a sample exactly parallel to one of the reference series, and the question was whether they would assign themselves

to that specific population. In this testing of New
Zealand Maoris, the sample is not drawn from
such a population. It is, however, a sample of
clearly known Polynesian derivation and, with
three Polynesian series among the base material,
the question is whether the Maori specimens
are recognized as generally Polynesian. This
broader kind of identification is what we might
expect, or hope for, in many other such queries
of tested skulls.

The 20 Maoris (all selected for measuring as
being probably male) are from varied localities:
10 from the Murihiku Coast of the South Island
and 10 from the North Island. Because there
exists no large collection of Maori crania well
localized to a specific area, Chatham Islands
Moriori were used as one of the reference series
for the original work (1989). The Moriori are
ultimately of the same derivation as the New
Zealand Maori, although it is not known
whether they arrived directly from central Polynesia or as an early migration from New Zealand
near the time of the original navigators. There
has evidently been mutual isolation by 500 miles
of ocean for a considerable period, but on the
face of it we might expect that Maori affiliations
found herein would be primarily with Moriori.

Table 3 shows the DISPOP results. In first
place—that is, closest to a Maori specimen—is
Easter Island in 9 cases, Moriori in 7 cases, and
Hawaii in 1 case. Of the remaining 3 cases:
Anyang is barely closer than Moriori for *#3026*;
#3021 is definitely closest to Arikara, but is well
within a Moriori perimeter at 12 percent probability; *#3018* is indistinguishably closer to Ainu
(places of decimals) but fairly distant from any
established centroid. The net result is a) good
allocation to "Polynesian" over all, and b) probability of wrong precise identification—too low a
P value—in less than 10 percent of cases. As
another aspect of identification, it is notable that
African or European groups are virtually absent
as possible assignments in the first ten positions.

This is pretty good evidence, remembering of
course the favoring circumstances, of: a) presence
among the reference series of likely populations
of affiliation, especially if a region is well represented, like Polynesia; and b) recency and nor-
mality of the target skull. It will be seen further
on that readings are less clear when a probable

TABLE 3

DISPOP results: Maori TEST cases

3016 Maori	Easter	Hawaii	Morior	Anyang	Ainu	Guam	N.Jap.	Hainan	S.Jap.	Atayal
	3.4	4.2	4.9	5.1	5.1	5.5	5.8	5.9	6.0	6.1
	0.64	0.20	0.04	0.02	0.02	0.01	0.00	0.00	0.00	0.00
3017 Maori	Hawaii	Peru	Morior	Easter	Arikar	Guam	N.Jap.	S.Cruz	Hainan	S.Jap.
	4.4	5.2	5.3	5.6	6.1	6.3	6.4	6.4	6.7	6.7
	0.14	0.01	0.01	0.00	0.00	0.00	-	-	-	-
3018 Maori	Ainu	Morior	Easter	Hawaii	Tasman	N.Jap.	Anyang	Tolai	Hainan	Atayal
	5.7	5.7	6.1	6.2	6.4	6.5	6.8	6.8	6.8	6.9
	0.00	0.00	0.00	0.00	-	-	-	-	-	-
3019 Maori	Easter	Anyang	Ainu	Morior	Hawaii	Guam	Hainan	N.Jap.	S.Jap.	Atayal
	4.7	4.9	5.0	5.1	5.2	5.4	5.6	5.8	5.9	6.0
	0.08	0.04	0.03	0.02	0.01	0.01	0.00	0.00	0.00	0.00
3020 Maori	Easter	Hawaii	Morior	Arikar	Peru	Tolai	S.Cruz	Guam	Tasman	Andamn
	4.0	4.1	4.3	5.8	5.9	6.0	6.0	6.3	6.4	6.8
	0.29	0.26	0.16	0.00	0.00	0.00	0.00	0.00	-	-
3021 Maori	Arikar	Morior	S.Cruz	Guam	Hawaii	Peru	Eskimo	Ainu	Zalvar	Norse
	3.7	4.5	4.7	5.2	5.9	6.0	6.1	6.1	6.2	6.3
	0.44	0.12	0.08	0.01	0.00	0.00	0.00	0.00	0.00	0.00
3022 Maori	Morior	Hawaii	Guam	Ainu	Anyang	N.Jap.	Arikar	Hainan	Easter	Peru
	4.4	4.7	5.2	5.7	5.9	5.9	6.0	6.1	6.2	6.2
	0.14	0.08	0.01	0.00	0.00	0.00	0.00	0.00	0.00	0.00
3023 Maori	Morior	Hawaii	Arikar	Guam	N.Jap.	S.Cruz	Easter	Anyang	Phil	Hainan
	3.3	4.3	4.9	5.1	5.7	5.7	5.8	5.8	6.1	6.1
	0.64	0.19	0.04	0.02	0.00	0.00	0.00	0.00	0.00	0.00
3024 Maori	Morior	Arikar	Hawaii	Guam	Easter	S.Cruz	Peru	Andamn	Anyang	Hainan
	4.7	5.1	5.4	5.7	5.8	5.9	6.1	6.3	6.4	6.4
	0.08	0.02	0.01	0.00	0.00	0.00	0.00	0.00	-	-
3063 Maori	Easter	Morior	Hawaii	Anyang	Guam	Hainan	S.Jap.	Atayal	Phil	Ainu
	5.3	5.9	6.0	6.6	7.4	7.4	7.6	7.6	7.7	7.8
	0.01	0.00	0.00	-	-	-	-	-	-	-
3025 N Maori	Easter	Hawaii	S.Jap.	Guam	Zulu	Morior	Phil	N.Jap.	Atayal	Hainan
	3.5	4.1	4.9	5.0	5.1	5.1	5.1	5.3	5.3	5.3
	0.54	0.28	0.04	0.03	0.03	0.02	0.02	0.01	0.01	0.01
3026 N Maori	Anyang	Morior	Easter	Hainan	Hawaii	Guam	Arikar	S.Jap.	Phil	N.Jap.
	4.3	4.3	4.8	4.8	4.9	4.9	5.4	5.4	5.5	5.6
	0.18	0.17	0.05	0.05	0.04	0.04	0.01	0.01	0.01	0.00
3027 N Maori	Morior	Tolai	Ainu	Tasman	Arikar	Hawaii	Guam	Easter	S.Cruz	N.Jap.
	3.9	4.4	4.6	4.7	4.7	4.7	4.8	4.8	4.9	5.0
	0.35	0.16	0.09	0.07	0.07	0.06	0.06	0.06	0.04	0.03
3028 N Maori	Easter	Hawaii	Morior	Tolai	Guam	Tasman	Arikar	S.Cruz	Eskimo	Ainu
	4.1	4.6	5.0	5.2	5.5	5.5	5.7	5.7	6.1	6.3
	0.28	0.09	0.03	0.02	0.01	0.00	0.00	0.00	0.00	0.00
3029 N Maori	Easter	Hawaii	Guam	Morior	Anyang	Arikar	Peru	S.Jap.	Hainan	N.Jap.
	3.7	4.1	4.2	4.3	4.8	4.8	4.9	5.0	5.2	5.4
	0.44	0.28	0.22	0.19	0.05	0.05	0.04	0.03	0.02	0.01
3030 N Maori	Morior	S.Cruz	Peru	Arikar	N.Jap.	Ainu	Norse	Hawaii	S.Jap.	Anyang
	3.0	4.4	5.1	5.3	5.3	5.4	5.5	6.0	6.0	6.2
	0.79	0.14	0.02	0.01	0.01	0.01	0.00	0.00	0.00	0.00
3031 N Maori	Easter	Morior	Tolai	Anyang	N.Jap.	S.Jap.	Hainan	Arikar	Atayal	Hawaii
	5.2	5.6	5.8	5.9	6.2	6.3	6.5	6.6	6.8	6.8
	0.01	0.00	0.00	0.00	0.00	0.00	-	-	-	-
3032 N Maori	Morior	S.Cruz	Hawaii	Tasman	Easter	Arikar	Ainu	Tolai	Austr	Guam
	4.7	5.7	5.9	6.0	6.0	6.0	6.1	6.2	6.5	6.6
	0.07	0.00	0.00	0.00	0.00	0.00	0.00	0.00	-	-
3033 N Maori	Morior	Ainu	Easter	N.Jap.	Hawaii	Tolai	Anyang	S.Jap.	Guam	Tasman
	4.7	5.7	5.9	6.1	6.5	6.7	6.9	6.9	7.2	7.2
	0.07	0.00	0.00	0.00	-	-	-	-	-	-
3064 N Maori	Easter	Ainu	Hawaii	Morior	Tasman	Tolai	Guam	Zalvar	Norse	Atayal
	4.8	5.7	5.8	5.9	6.1	6.5	7.2	7.4	7.5	7.5
	0.05	0.00	0.00	0.00	0.00	-	-	-	-	-

major region of derivation is not represented, or poorly so, in the basic information matrix, or else when a target skull is not recent in origin.

Professor van Vark has applied POSCON to these Maori cases. Results are comparable without coinciding, and in fact very satisfying. The diagnosis is 12 Moriori, 3 Easter Island, 2 Hawaii, 2 Arikara, and 1 Guam, again giving only 3 non-Polynesian, and more Moriori, as is probably to be expected. Actual coincidence is less than this suggests. Van Vark's "Arikara" (*#3024*, *#3032*) are my "Moriori" while my "Arikara" (*#3021*) or "Ainu" (*#3018*) are his "Moriori", and so on. The lesson seems to be that the two methods, differing in detail though similar in approach, are effective in general placement of this group of skulls.

3 Trials on Known Cases

We may test the methods better by applying them to other members of the TEST series. These are specimens generally outside the control populations, but possibly related to them regionally or ethnically, and thus candidates for affiliation. The TEST specimens were not collected systematically (see Howells 1989) but rather as targets of opportunity, as we traveled from institution to institution as necessary to collect data on the main populations. They were nonetheless measured with the idea of later examinations like the present one. The members of the whole TEST series are preponderantly "male."

This exercise is not simply a test of method. As will be seen, there is certainly anthropological information involved, which is one object of the whole investigation.

I begin by applying DISPOP to individuals in the TEST series, mainly assembled in small lots, although members of the groups shown are not always from the same collections. They are listed by region in the many parts of Table 4 below. In this, both males and females are shown. Only the 8 nearest populations by D are given, and female cases are offset to the right for identification. (In cases where close examination is indicated, POPKIN was also applied, with results given below in the relevant section.)

Europe

The TEST series is deficient in "European" skulls which, in institutions housing them, tend to occur in major groups or collections, rather than as small units with a strongly local identification. (See Howells 1973, 1989 for criteria of the selection of the main European series as representing identifiable populations.)

Nine Finnish Lapps, hardly typical Europeans—and for whom we might not expect a parent population—nevertheless are read as European, with a dash of "Buriat," which may express short-headedness as much as ethnic affinity. "Berg" as another possibility may signify the same.

The Avars were horse-riding warrior nomads from Central Asia who set up an empire from Austria east and south, between the Roman withdrawal and the arrival of the Magyar "Conquering Hungarians," an occupation from the sixth to the eighth centuries. The remnant Avars merged with other settlers, new and old.

Ferencz (1991) compared crania from 7 Avar cemeteries, finding the southern (Yugoslavian) group to exhibit Mongoloid characters while the northern (Hungarian) group had a "strong Europoid character." A large sample from Lower

TABLE 4-1
DISPOP results: Europe

1549 Lapp	Berg	Buriat	Norse	Zalvar	S.Cruz	Bush	N.Jap.	Peru
	4.2	4.8	5.1	5.3	5.8	6.3	6.4	6.5
	0.20	0.05	0.02	0.01	0.00	0.00	-	-
1550 Lapp	Buriat	Berg	Zalvar	Peru	Arikar	Norse	S.Cruz	Guam
	6.1	6.3	7.2	7.3	7.3	7.5	7.5	8.1
	0.00	0.00	-	-	-	-	-	-
1551 Lapp	Zalvar	Berg	Andamn	Egypt	Norse	Phil	Atayal	Hainan
	4.4	5.0	5.2	5.2	5.3	5.4	5.5	5.6
	0.13	0.03	0.02	0.01	0.01	0.01	0.01	0.00
1556 Lapp	Zalvar	Norse	Berg	Teita	Egypt	S.Cruz	Eskimo	Arikar
	4.4	4.6	5.3	5.3	5.3	5.3	5.7	5.9
	0.16	0.09	0.01	0.01	0.01	0.01	0.00	0.00

1552 Lapp	Berg	Arikar	Zalvar	Morior	Ainu	Norse	Tasman	Atayal
	4.4	4.5	5.0	5.3	5.4	5.5	5.5	5.9
	0.13	0.12	0.02	0.01	0.01	0.00	0.00	0.00
1553 Lapp	Berg	Zalvar	Arikar	Buriat	Norse	Atayal	Ainu	N.Jap.
	3.9	5.2	5.7	6.3	6.4	6.6	6.8	6.9
	0.35	0.01	0.00	-	-	-	-	-
1554 Lapp	Norse	Arikar	Zalvar	Egypt	Berg	Buriat	S.Cruz	N.Jap.
	4.4	5.1	5.1	5.2	5.2	5.8	5.8	6.0
	0.14	0.02	0.02	0.01	0.01	0.00	0.00	0.00
1555 Lapp	Buriat	Berg	Arikar	Hainan	Zalvar	Atayal	N.Jap.	S.Jap.
	3.9	4.3	4.4	4.9	5.0	5.1	5.1	5.3
	0.37	0.19	0.14	0.03	0.03	0.02	0.02	0.01
1557 Lapp	Berg	Zalvar	Atayal	Egypt	N.Jap.	Andamn	Norse	S.Jap.
	4.7	5.0	5.0	5.0	5.1	5.1	5.2	5.3
	0.07	0.03	0.03	0.02	0.02	0.02	0.01	0.01

1779 Canary Is.	Zalvar	Egypt	Norse	Ainu	Berg	Austr	Guam	Phil
	3.5	3.7	3.9	4.5	4.9	5.3	5.8	5.9
	0.55	0.45	0.36	0.11	0.04	0.01	0.00	0.00
1780 Canary Is.	Norse	Egypt	Zalvar	Berg	S.Cruz	Peru	Arikar	Ainu
	3.9	4.8	4.9	5.3	5.5	5.9	6.1	6.1
	0.36	0.06	0.05	0.01	0.01	0.00	0.00	0.00
1781 Canary Is.	Norse	Berg	Zalvar	Ainu	Egypt	Atayal	Peru	Phil
	3.2	3.5	3.5	4.0	4.3	5.1	5.1	5.2
	0.72	0.57	0.54	0.30	0.18	0.03	0.02	0.02
1565 Avar	Buriat	Berg	Eskimo	Guam	Zalvar	Norse	Ainu	Arikar
	4.1	7.6	7.7	8.0	8.0	8.2	8.6	8.7
	0.25	-	-	-	-	-	-	-
1566 Avar	Buriat	Teita	Norse	Anyang	Hainan	S.Jap.	Zalvar	Atayal
	5.3	5.7	5.8	6.1	6.1	6.3	6.3	6.6
	0.01	0.00	0.00	0.00	0.00	0.00	0.00	-
1567 Avar	Buriat	Eskimo	S.Cruz	N.Jap.	Zalvar	Arikar	Guam	Norse
	5.2	6.2	6.3	6.6	6.6	6.7	6.7	6.8
	0.02	0.00	0.00	-	-	-	-	-
1568 Avar	S.Cruz	Zalvar	Arikar	Norse	Peru	Austr	N.Jap.	Ainu
	4.0	4.5	4.5	4.6	4.7	4.8	4.9	4.9
	0.29	0.12	0.11	0.08	0.07	0.05	0.04	0.04

1569 Avar	Norse	Arikar	Peru	Egypt	Zalvar	Berg	S.Cruz	Andamn
	3.8	4.5	4.6	4.7	4.8	4.8	4.9	6.3
	0.38	0.10	0.07	0.06	0.05	0.05	0.03	-
1570 Avar	Berg	Arikar	Zalvar	Peru	S.Cruz	N.Jap.	S.Jap.	Buriat
	4.9	5.1	5.3	5.3	5.6	5.7	5.8	5.9
	0.03	0.02	0.01	0.01	0.00	0.00	0.00	0.00

Austria (Zwölfaxing) studied by Kritscher and Szilvássy (1988/89) also exhibited a primarily European impression (*Gepräge*) as well as measurement means departing from various Inner Asiatic peoples.

The male Avars here, from Hungarian localities, do not suggest Europeans. As far as they approach any peoples, it is the Siberian Buriats. Females, however, would support the views of the above authors.

Guanches (N = 3) of the Canary Islands are allocated as clearly "European," their traditional affiliation (see Hooton 1925), with no suggestion of any sub-Saharan connections.

Africa

Nubians are Egyptian/European in connections. With a couple of exceptions, other Africans shown are strongly "African" in affiliation, with Dogon or Zulu prevailing over Teita (a Kenyan Bantu tribe) as the most likely connection.

The Bantu-speaking Haya inhabit the northwest corner of Tanzania, so their likeness to Zulus is reasonable in spite of their geographical proximity to Teita, also Bantu-speakers. (The skulls themselves come from burial caves on an island in Lake Victoria.) One female approaches Tolais of Melanesia, not very closely.

Jebel Moya, in the Sudan, is about 250 km south southeast of Khartoum. The population dates from 1000 B.C. on and is represented by a large cemetery. Mukherjee, Rao, and Trevor (1955) used D^2 to relate the series to other Africans, finding it closest to West African groups (Ibo, Cameroons) and furthest from such as Tigre, Egyptian E (our Egyptian reference series) but also Teita. Results for the two males here agree regarding West Africans being closer than Teita, though the latter are by no means distant absolutely. The lone female comes out as "European."

West African Dahomeans appropriately come out as Dogon or Zulu. (Two cases are very remote from all populations, though Africans are the least remote.) The West African Pygmy cases also appear most often as closest to Dogons. In no case of Pygmies is there any approach to South African Bushmen; but Andamanese Negritos appear in the offing, nosing out Dogon in the case of one female.

TABLE 4-2

DISPOP results: Africa

1579 Nubia	Zalvar	Norse	Egypt	Berg	S.Cruz	Arikar	Phil	Ainu
	2.4	2.9	3.7	3.7	3.8	4.4	4.6	4.7
	0.95	0.82	0.47	0.46	0.40	0.13	0.09	0.06
1580 Nubia	Teita	Egypt	Zalvar	Zulu	Norse	Phil	Guam	Andamn
	2.7	4.0	4.2	4.3	4.3	4.3	4.4	4.6
	0.89	0.30	0.21	0.17	0.17	0.16	0.15	0.10
1583 Nubia	Norse	Teita	Zalvar	Egypt	Austr	Zulu	Berg	Tolai
	4.6	4.7	4.7	5.3	5.3	5.5	5.6	6.0
	0.09	0.08	0.07	0.01	0.01	0.01	0.00	0.00
1581 Nubia	Egypt	Norse	Zalvar	Andamn	Tolai	S.Jap.	N.Jap.	Atayal
	4.6	5.2	5.2	5.2	5.4	5.6	5.6	5.7
	0.08	0.01	0.01	0.01	0.01	0.00	0.00	0.00
1582 Nubia	Egypt	Norse	Zalvar	N.Jap.	Zulu	Ainu	Teita	S.Jap.
	4.6	4.8	4.8	5.3	5.4	5.5	5.5	5.6
	0.09	0.05	0.05	0.01	0.01	0.00	0.00	0.00
1584 Nubia	Egypt	Atayal	Zalvar	Zulu	Bush	Norse	Ainu	Andamn
	4.1	4.3	4.4	4.6	4.8	4.8	4.9	4.9
	0.25	0.16	0.15	0.09	0.05	0.04	0.03	0.03
1589 Haya Tribe	Zulu	Dogon	Ainu	Phil	Atayal	Andamn	Bush	Egypt
	3.2	3.5	4.7	5.0	5.0	5.2	5.3	5.4
	0.71	0.56	0.08	0.03	0.03	0.01	0.01	0.01
1590 Haya Tribe	Zulu	Dogon	Teita	N.Jap.	S.Jap.	Phil	Atayal	Egypt
	4.1	4.6	5.6	6.5	6.5	6.8	6.9	6.9
	0.27	0.10	0.00	-	-	-	-	-

Who's Who in Skulls

T ABLE 4-2 CONTINUED

DISPOP results: Africa

1591 Haya Tribe	Dogon	Zulu	Teita	Ainu	Egypt	Andamn	Atayal	Tolai	
	4.1	4.7	5.1	7.1	7.1	7.4	7.5	7.5	
	0.24	0.07	0.02	-	-	-	-	-	
1592 Haya Tribe	Tolai	Zulu	Austr	S.Cruz	Arikar	S.Jap.	Teita	Ainu
	4.9	5.1	6.3	6.6	6.9	6.9	6.9	7.0
	0.03	0.02	-	-	-	-	-	-
1593 Haya Tribe	Teita	Tolai	Bush	Zulu	Dogon	Austr	Egypt	Norse
	4.3	5.5	5.8	6.1	6.5	6.6	6.8	6.9
	0.17	0.00	0.00	0.00	-	-	-	-
1594 Haya Tribe	Zulu	Atayal	Teita	Zalvar	Ainu	S.Jap.	Dogon	Tolai
	4.3	4.8	4.9	5.3	5.3	5.7	5.7	5.8
	0.18	0.05	0.03	0.01	0.01	0.00	0.00	0.00
1585 Jebel Moya	Dogon	Teita	Zulu	S.Jap.	Phil	Atayal	Andamn	Hainan
	4.2	4.3	4.4	5.0	5.7	5.7	5.8	5.8
	0.22	0.17	0.15	0.03	0.00	0.00	0.00	0.00
1586 Jebel Moya	Dogon	Teita	Andamn	Zulu	Phil	Hainan	Anyang	S.Jap.
	4.0	4.6	5.2	5.5	5.8	6.0	6.2	6.2
	0.29	0.09	0.02	0.01	0.00	0.00	0.00	0.00
1587 Jebel Moya	Zalvar	Norse	Egypt	Atayal	Berg	Ainu	S.Jap.	Arikar
	4.0	4.7	5.0	5.0	5.4	5.7	5.7	5.9
	0.32	0.06	0.02	0.02	0.01	0.00	0.00	0.00
1596 Dahomey	Zulu	Dogon	Egypt	Teita	Ainu	S.Jap.	Phil	Zalvar
	4.3	4.3	5.0	5.5	5.7	6.1	6.2	6.3
	0.18	0.18	0.03	0.01	0.00	0.00	0.00	0.00
1597 Dahomey	Dogon	Zulu	Austr	Egypt	Tolai	Teita	Ainu	Zalvar
	6.7	7.0	7.1	7.6	7.7	7.8	7.9	8.2
	-	-	-	-	-	-	-	-
1598 Dahomey	Zulu	Teita	Dogon	Tolai	Guam	Egypt	Austr	Atayal
	4.0	4.7	5.7	6.4	6.6	6.8	6.8	6.9
	0.32	0.06	0.00	-	-	-	-	-
1595 Dahomey	Zulu	Dogon	Teita	Tolai	Austr	Bush	Ainu	Atayal
	5.7	6.0	6.3	7.0	7.5	7.7	7.8	7.9
	0.00	0.00	-	-	-	-	-	-
1599 Dahomey	Dogon	Teita	Zulu	Ainu	Atayal	Tolai	Egypt	S.Jap.
	3.6	4.6	4.7	5.4	5.5	5.6	5.7	5.8
	0.54	0.09	0.06	0.01	0.00	0.00	0.00	0.00
1600 Dahomey	Dogon	Zulu	Bush	Teita	Atayal	Andamn	Egypt	Ainu
	3.3	4.4	5.0	5.3	5.5	5.7	5.7	5.9
	0.66	0.12	0.03	0.01	0.00	0.00	0.00	0.00
1601 Pygmy	Teita	Dogon	Andamn	Zulu	Egypt	Tolai	Tasman	Austr
	4.9	5.1	5.8	5.8	5.8	6.7	7.0	7.0
	0.05	0.02	0.00	0.00	0.00	-	-	-
1602 Pygmy	Dogon	Zulu	Teita	Phil	S.Jap.	Bush	Andamn	Tolai
	3.5	4.6	5.2	5.6	5.9	6.0	6.0	6.1
	0.55	0.09	0.02	0.00	0.00	0.00	0.00	0.00
1603 Pygmy	Dogon	Zulu	Teita	S.Jap.	Atayal	Andamn	Egypt	Ainu
	4.2	5.1	5.3	6.7	6.8	6.9	6.9	7.0
	0.21	0.02	0.01	-	-	-	-	-
1604 Pygmy	Dogon	Andamn	Teita	Tolai	Zulu	Austr	Bush	Atayal
	4.5	4.9	5.1	5.7	6.1	6.8	6.8	6.9
	0.10	0.03	0.02	0.00	0.00	-	-	-
1605 Pygmy	Andamn	Dogon	Egypt	Zulu	Atayal	Hainan	Peru	Zalvar
	4.1	4.2	4.2	4.8	5.3	5.3	5.3	5.3
	0.24	0.21	0.20	0.05	0.01	0.01	0.01	0.01

Australia

Of seven Australians, five from the Northern Territory appear closer to Tasmanians than to South Australians (for whom the reference population is from the Lower Murray River region). This unsurprising result can be used in discussions of Australian populations, for which this is not the place. In five of these, Melanesian Tolais are a possible assignment at low levels of probability and this is the only possibility for a Queensland specimen. (Many Solomon Islanders were introduced into Queensland for plantation labor in the last century; we might have caught one here.)

Melanesia

For the Bismarck Archipelago and New Guinea, some of the assignments are only broadly satisfying. One Baining of New Britain and two New Guinea skulls have good probabilities of assignment as Tolai (New Britain, neighbors of the Baining) or Tasmanian. But the New Irelander is "wrong."

The Solomon Islands examples are drawn from Guadalcanal, Malaita, Isabel, and Choiseul. Males are appropriately diagnosed as Tolai with

a couple of exceptions; *#3056* ("South Japan") may be an immature individual. The two cases from the easterly Santa Cruz group are safely "Tolai," though Teita comes threateningly close for the female individual.

Considering the apparent physical variation in the living people within the New Hebrides (Vanuatu), as in Melanesia generally, the predilection for a Tolai affiliation is interesting, although 4 out of 12 are inappropriately read. New Caledonia and the Loyalty Islands come out better; the reading of "Guam" for *#2049*, a Lifu female, is perhaps not too anomalous, because of an evident non-Melanesian contribution to the Loyalties (Polynesian "outlier"?).

Altogether these Melanesian examples make a good case for successful assignment to a region.

Polynesia

For Central Polynesia, diagnoses are commonly "Hawaii," with a few for Easter and Moriori. "Guam" for one Tongan is not very bad. The extreme distances for Tongan *#3047* and *#3048* may be related to occipital flattening, common in Tonga. The Samoan and Hervey Island specimens are clearly misclassified. But on the whole classification to general area is distinctly good.

TABLE 4-3
DISPOP results: Australia

1634 N.Territory	Austr	Tasman	S.Cruz	Tolai	Peru	Arikar	Zalvar	Phil
	5.0	5.6	6.4	6.6	6.7	7.0	7.0	7.4
	0.03	0.00	-	-	-	-	-	-
1635 N.Territory	Tasman	Austr	Tolai	Zulu	Teita	Bush	Andamn	Zalvar
	3.3	4.0	4.8	6.0	6.1	6.1	6.3	6.4
	0.64	0.31	0.05	0.00	0.00	0.00	0.00	-
1636 N.Territory	Tasman	Tolai	Austr	Easter	Zulu	Guam	S.Cruz	Ainu
	4.4	4.6	5.4	7.7	8.1	8.2	8.3	8.4
	0.14	0.09	0.01	-	-	-	-	-
1637 N.Territory	Tasman	Austr	Tolai	Atayal	Ainu	Zulu	Zalvar	N.Jap.
	3.6	4.1	4.8	5.2	5.5	5.8	6.0	6.0
	0.53	0.23	0.05	0.01	0.00	0.00	0.00	0.00
1638 N.Territory	Tasman	Tolai	Austr	Zulu	Atayal	Andamn	Zalvar	Dogon
	4.2	5.0	5.1	5.3	5.3	5.4	5.4	5.5
	0.22	0.03	0.02	0.01	0.01	0.01	0.01	0.00
1639 N.Territory	Tasman	Tolai	Austr	Andamn	Atayal	S.Cruz	Arikar	Zalvar
	4.0	4.3	4.4	5.7	6.0	6.2	6.2	6.3
	0.31	0.19	0.12	0.00	0.00	0.00	0.00	-
1640 Cairns Dist.	Tolai	Tasman	Bush	Austr	Atayal	Easter	Andamn	S.Cruz
	4.8	5.6	5.9	5.9	6.1	6.2	6.4	6.4
	0.04	0.00	0.00	0.00	0.00	0.00	-	-

Who's Who in Skulls

TABLE 4-4
DISPOP results: Melanesia

1644 Baining	Tasman	Phil	Tolai	Guam	Arikar	Hainan	N.Jap.	S.Jap.	
	5.3	5.5	5.7	6.0	6.0	6.2	6.5	6.5	
	0.01	0.00	0.00	0.00	0.00	0.00	-	-	
1645 Baining	Tolai	Hawaii	Tasman	Ainu	Austr	Morior	Arikar	Atayal	
	4.0	4.4	4.5	5.1	5.1	5.2	5.3	5.4	
	0.28	0.12	0.10	0.02	0.02	0.01	0.01	0.01	
3058 New Ireland	Peru	Hawaii	Tolai	Phil	Arikar	Morior	S.Cruz	Hainan	
	4.8	5.0	5.4	5.5	5.6	5.6	5.7	5.8	
	0.06	0.03	0.01	0.00	0.00	0.00	0.00	0.00	
3059 New Guinea	Austr	Tasman	Tolai	Zulu	Zalvar	Bush	Teita	S.Cruz	
	5.1	5.2	5.5	6.2	6.6	6.9	7.0	7.1	
	0.02	0.01	0.01	0.00	-	-	-	-	
3060 New Guinea	Tolai	Tasman	S.Cruz	Andamn	Peru	Austr	Teita	Atayal	
	4.2	5.2	5.5	5.8	6.1	6.2	6.2	6.4	
	0.21	0.02	0.01	0.00	0.00	0.00	0.00	-	
3061 New Guinea	Tasman	Tolai	Andamn	Austr	Egypt	S.Cruz	Zalvar	Norse	
	3.6	3.7	5.2	5.2	5.2	5.3	5.3	5.4	
	0.50	0.45	0.02	0.02	0.01	0.01	0.01	0.01	
2024 New Guinea	Tolai	Tasman	Zulu	Easter	Hawaii	Morior	Egypt	Austr	
	5.8	6.2	6.3	6.5	6.9	7.0	7.1	7.1	
	0.00	0.00	-	-	-	-	-	-	
1647 Solomon Is.	Tolai	Phil	Arikar	S.Cruz	Zalvar	Tasman	Andamn	Hainan	
	4.8	4.9	5.1	5.1	5.2	5.3	5.3	5.4	
	0.05	0.04	0.03	0.02	0.01	0.01	0.01	0.01	
2025 Solomon Is.	Tolai	Phil	Tasman	N.Jap.	Zulu	S.Cruz	Dogon	Hainan	
	5.7	5.8	5.8	5.9	6.1	6.4	6.4	6.7	
	0.00	0.00	0.00	0.00	0.00	-	-	-	
2026 Solomon Is.	Tolai	N.Jap.	Tasman	S.Jap.	Phil	Hainan	Atayal	Guam	
	3.7	4.6	4.8	4.9	4.9	5.0	5.1	5.1	
	0.48	0.10	0.05	0.04	0.04	0.03	0.02	0.02	
2027 Solomon Is.	Tolai	Tasman	Austr	Peru	Morior	S.Cruz	Andamn	Ainu	
	4.4	5.2	5.7	6.0	6.0	6.3	6.6	6.7	
	0.15	0.01	0.00	0.00	0.00	0.00	-	-	
2028 Solomon Is.	Tolai	Guam	Phil	S.Jap.	N.Jap.	Atayal	Peru	Anyang	
	4.1	5.1	5.7	5.8	5.9	5.9	5.9	6.0	
	0.27	0.02	0.00	0.00	0.00	0.00	0.00	0.00	
2029 Solomon Is.	Tolai	Zulu	Andamn	Peru	Tasman	S.Cruz	Easter	Teita	
	4.6	4.8	5.3	5.7	5.7	5.7	5.8	5.8	
	0.10	0.06	0.01	0.00	0.00	0.00	0.00	0.00	
2030 Solomon Is.	S.Cruz	Berg	Peru	Arikar	Zalvar	Phil	Tolai	Tasman	
	4.5	4.8	4.8	4.9	5.0	5.0	5.1	5.3	
	0.11	0.06	0.06	0.04	0.03	0.03	0.02	0.01	
2031 Solomon Is.	Tasman	S.Cruz	Berg	Zalvar	Hawaii	Tolai	Norse	Arikar	
	4.6	5.9	5.9	6.1	6.2	6.2	6.4	6.4	
	0.08	0.00	0.00	0.00	0.00	0.00	-	-	
2719 Solomon Is.	Tolai	Tasman	Ainu	Easter	N.Jap.	Zalvar	Eskimo	S.Jap.	
	4.9	5.7	5.9	6.1	6.2	6.3	6.5	6.6	
	0.04	0.00	0.00	0.00	0.00	0.00	-	-	
2893 Solomon Is.	Tolai	Tasman	Arikar	Morior	Guam	S.Cruz	Austr	Ainu	
	4.7	6.0	6.0	6.1	6.2	6.3	6.4	6.6	
	0.08	0.00	0.00	0.00	0.00	0.00	-	-	
2894 Solomon Is.	Andamn	Tolai	Egypt	Teita	Norse	Zalvar	S.Cruz	Ainu	
	4.8	4.9	4.9	5.1	5.2	5.2	5.2	5.3	
	0.05	0.04	0.04	0.03	0.02	0.02	0.01	0.01	
2895 Solomon Is.	Tasman	Tolai	Zulu	Phil	Hawaii	Easter	Austr	Bush	
	3.6	4.5	5.0	5.2	5.4	5.5	5.5	5.5	
	0.48	0.11	0.03	0.01	0.01	0.01	0.01	0.01	

TABLE 4-4 CONTINUED

DISPOP results: Melanesia

2896 Solomon Is.	Tolai	Tasman	Austr	S.Cruz	Ainu	Easter	Atayal	Peru
	3.2	4.1	5.3	5.5	5.7	5.7	5.9	6.0
	0.73	0.24	0.01	0.01	0.00	0.00	0.00	0.00
3056 Solomon Is.	S.Jap.	N.Jap.	Andamn	Hainan	Tolai	Phil	S.Cruz	Atayal
	3.9	4.1	4.1	4.5	4.6	4.6	4.6	4.6
	0.35	0.25	0.25	0.11	0.10	0.10	0.09	0.08
2033 Solomon Is.	Tolai	Easter	Austr	Zulu	N.Jap.	Atayal	Teita	S.Jap.
	4.3	5.2	5.4	5.4	5.4	5.5	5.5	5.6
	0.16	0.01	0.01	0.01	0.00	0.00	0.00	0.00
2032 Solomon Is.	Zalvar	N.Jap.	Atayal	Ainu	Norse	S.Jap.	Tolai	Egypt
	3.3	3.5	3.5	3.5	3.7	3.8	4.0	4.0
	0.70	0.59	0.57	0.57	0.47	0.42	0.31	0.28
2720 Solomon Is.	Andamn	Atayal	S.Jap.	N.Jap.	S.Cruz	Ainu	Arikar	Morior
	4.1	4.2	4.2	4.3	4.4	4.4	4.7	4.7
	0.27	0.22	0.19	0.17	0.14	0.12	0.07	0.06
2721 Solomon Is.	Zulu	Norse	Tolai	Austr	Zalvar	Egypt	S.Cruz	Peru
	4.4	4.5	4.6	4.7	4.7	4.9	5.1	5.2
	0.14	0.11	0.08	0.07	0.05	0.03	0.02	0.01
3057 Solomon Is.	Zulu	Austr	Tolai	Dogon	Teita	Egypt	Zalvar	Atayal
	4.5	4.7	4.9	5.0	5.2	5.5	5.7	6.0
	0.10	0.06	0.04	0.03	0.01	0.00	0.00	0.00
2034 Santa Cruz I	Tolai	Austr	Zalvar	Guam	Tasman	Teita	S.Jap.	N.Jap.
	3.9	5.6	5.8	5.9	6.0	6.1	6.1	6.1
	0.34	0.00	0.00	0.00	0.00	0.00	0.00	
2035 Santa Cruz I.	Tolai	Teita	Zulu	Easter	Atayal	Austr	Dogon	S.Jap.
	3.5	3.5	4.8	4.9	4.9	5.1	5.2	5.3
	0.59	0.55	0.05	0.04	0.03	0.02	0.01	0.01
2036 Vanuatu	Zulu	Tolai	Austr	Zalvar	Atayal	Teita	Peru	Tasman
	4.8	4.8	5.6	5.7	6.3	6.3	6.3	6.3
	0.05	0.05	0.00	0.00	0.00	0.00	0.00	0.00
2037 Vanuatu	Tolai	Austr	Tasman	Teita	Zulu	S.Cruz	Andamn	Zalvar
	3.0	3.6	4.0	4.7	5.0	5.1	5.4	5.5
	0.78	0.51	0.30	0.07	0.03	0.03	0.01	0.00
2040 Vanuatu	Tolai	Austr	Teita	Tasman	Dogon	Norse	Egypt	Zulu
	5.1	6.5	6.8	7.0	7.4	7.6	7.6	7.8
	0.02	-	-	-	-	-	-	-
2041 Vanuatu	Tolai	Zulu	Tasman	Egypt	Peru	Easter	Zalvar	Dogon
	4.3	5.6	5.6	5.9	6.1	6.3	6.4	6.5
	0.16	0.00	0.00	0.00	0.00	0.00	-	-
2042 Vanuatu	Tolai	Tasman	Austr	Peru	Guam	Teita	Phil	Easter
	3.3	5.2	5.4	5.4	5.5	5.6	5.7	5.7
	0.67	0.01	0.01	0.01	0.01	0.00	0.00	0.00
2043 Vanuatu	Phil	Andamn	Tasman	Guam	Hainan	Hawaii	Tolai	Anyang
	4.1	4.1	4.8	4.8	5.0	5.1	5.1	5.3
	0.26	0.24	0.06	0.05	0.03	0.02	0.02	0.01
2044 Vanuatu	Tolai	Tasman	Austr	Zulu	Egypt	S.Cruz	Zalvar	Dogon
	4.1	4.8	5.6	5.8	6.4	6.6	6.7	6.7
	0.27	0.05	0.00	0.00	-	-	-	-
2038 Vanuatu	Easter	Tolai	Tasman	Atayal	Morior	Zalvar	Ainu	Zulu
	4.1	4.8	5.1	5.7	6.0	6.0	6.2	6.3
	0.23	0.04	0.01	0.00	0.00	0.00	-	-
2039 Vanuatu	Tolai	Tasman	Austr	Zulu	Easter	Atayal	Dogon	Hawaii
	4.0	4.9	5.2	5.3	5.7	5.8	6.1	6.4
	0.31	0.04	0.01	0.01	0.00	0.00	0.00	-
2045 Vanuatu	Tolai	Easter	Austr	Zulu	Tasman	N.Jap.	Atayal	S.Jap.
	4.2	5.9	6.0	6.1	6.2	6.4	6.5	6.6
	0.19	0.00	0.00	0.00	0.00	-	-	-

TABLE 4-4 CONTINUED
DISPOP results: Melanesia

2046 Vanuatu	Tolai	Easter	Zulu	Austr	Tasman	Teita	Morior	Atayal
	3.2	4.6	5.1	5.3	5.5	5.5	5.8	5.8
	0.72	0.07	0.02	0.01	0.00	0.00	0.00	0.00

3055 Vanuatu	Zulu	Dogon	Andamn	Atayal	Hainan	S.Jap.	Guam	Tolai
	3.4	3.9	4.2	4.3	4.4	4.6	4.8	4.8
	0.65	0.35	0.20	0.19	0.13	0.07	0.05	0.05

2047 N. Caledonia	Tolai	Peru	S.Cruz	Tasman	Austr	Zulu	Atayal	Andamn
	4.5	5.3	5.5	5.6	5.7	5.9	6.0	6.1
	0.12	0.01	0.01	0.00	0.00	0.00	0.00	0.00

2048 N. Caledonia	Tolai	Zulu	Egypt	Teita	Zalvar	Tasman	Easter	Andamn
	3.5	4.8	5.0	5.0	5.2	5.3	5.4	5.5
	0.59	0.05	0.03	0.03	0.02	0.01	0.01	0.01

1648 N. Caledonia	Tolai	Hawaii	Easter	Guam	N.Jap.	S.Jap.	Peru	Zulu
	4.8	5.3	5.6	6.1	6.5	6.7	6.7	6.8
	0.06	0.01	0.00	0.00	-	-	-	-

2049 Loyalty Is.	Guam	S.Jap.	Anyang	N.Jap.	Tolai	Hainan	Easter	Eskimo
	3.9	4.8	4.9	4.9	5.0	5.3	5.5	5.6
	0.34	0.05	0.05	0.05	0.03	0.01	0.01	0.00

2050 Loyalty Is.	Tolai	Easter	Peru	Guam	Tasman	Anyang	Morior	Atayal
	5.1	5.3	6.3	6.6	6.8	7.0	7.1	7.1
	0.02	0.01	0.00	-	-	-	-	-

Micronesia

For the Marianas, the Guam specimen is not related to the main series and has an anomalous shape for that sample. The skulls from Tinian and Saipan are satisfyingly identified with Guam. That from Rota is read as Ainu, with Guam close behind. The Guam and Ainu main series are apt to cluster together in runs shown in *Skull Shapes*.

Eastern Micronesia is more confusing, with assignments scattered widely. Polynesian affiliation is rebuffed in any case. In some clustering studies (e.g., Howells 1970, on living subjects) Micronesian affiliations with Melanesia have been strongly suggested in spite of appearances. Still, Andamanese connections seem off the mark, and connections to Egypt or Teita (remote) are certainly not helpful. The answer may lie in ignorance. I do not know the history of the individual skulls, but during the last century these small populations were exposed to non-Oceanic ship jumpers of all kinds, American Blacks included, as can be documented for some islands like Nauru. So attempts to explain some assignments entirely in terms of aboriginal populations here might be a mistake. In any case, central Micronesia is not really represented in the reference populations—the Marianas, with Guam, tend to separate cranially from the rest of that area (see Howells n.d.)

Boreal Peoples

All Eskimos, whether Greenland (non-Inugsuk), Point Barrow, or Siberia, have "Eskimo" (here meaning Greenland, Inugsuk culture) as first affiliation, although at least one Siberian Eskimo is extremely remote in probability, with an SD distance of 6.4. The cases labeled "Central Arctic" were originally retrieved in the 1960s, and sent to Carleton Coon with the suggestion that they were either European or admixed with Caucasoids. Coon lent them to me for measurement. All are diagnosed, with varying degrees of probability, as "Eskimo." Among all these crania, ethnic identification is excellent.

A lone Aleut lines up instead with American Indians, which accords entirely with analyses by others, especially Nancy Ossenberg (1994).

At the opposite side of the Bering Strait, three male Chukchis line up as Eskimo. One female agrees, while two others affiliate, at low levels of probability, with other Asiatic populations.

TABLE 4-5

DISPOP results: Polynesia

2685 Tonga	Guam	Hawaii	Morior	Ainu	Arikar	Tasman	Tolai	Phil	
	5.0	5.1	5.2	5.3	5.6	5.9	6.1	6.2	
	0.03	0.02	0.02	0.01	0.00	0.00	0.00	0.00	
2686 Tonga	Hawaii	Phil	Hainan	Guam	Tolai	N.Jap.	Easter	Anyang	
	5.0	5.6	5.9	5.9	6.0	6.0	6.2	6.3	
	0.03	0.00	0.00	0.00	0.00	0.00	0.00	0.00	
3048 Tonga	Hawaii	Arikar	Berg	Morior	Peru	Guam	Norse	Hainan	
	6.6	7.0	7.2	7.3	7.5	7.9	8.0	8.0	
	-	-	-	-	-	-	-	-	
3049 Tonga	Easter	Morior	Hawaii	Tolai	Guam	Ainu	Hainan	Anyang	
	4.6	5.1	5.2	5.6	5.7	5.7	5.9	5.9	
	0.10	0.02	0.02	0.00	0.00	0.00	0.00	0.00	
3050 Tonga	Hawaii	Easter	Morior	Guam	Phil	Hainan	Anyang	Arikar	
	4.9	6.1	6.6	7.0	7.1	7.1	7.1	7.5	
	0.04	0.00	-	-	-	-	-	-	
2897 Tonga	N.Jap.	S.Jap.	Anyang	Morior	Hainan	Easter	Hawaii	Phil	
	5.0	5.4	5.4	5.7	6.0	6.1	6.1	6.1	
	0.03	0.01	0.01	0.00	0.00	0.00	0.00	0.00	
2687 Tonga	Hawaii	Hainan	S.Jap.	Guam	Andamn	Arikar	Peru	Ainu
	3.9	4.6	5.5	5.6	5.7	5.7	5.8	5.9
	0.34	0.08	0.00	0.00	0.00	0.00	0.00	0.00
3047 Tonga	Buriat	Arikar	Berg	Hainan	S.Jap.	Atayal	Hawaii	Andamn
	6.6	6.6	6.9	7.3	7.8	7.9	7.9	8.0
	-	-	-	-	-	-	-	-
2898 Samoa	Hainan	Andamn	Phil	Guam	Anyang	Atayal	Peru	Arikar	
	4.0	4.0	4.1	4.4	4.6	4.7	4.8	4.9	
	0.29	0.28	0.27	0.13	0.09	0.07	0.05	0.04	
1649 Tahiti	Easter	Hawaii	Guam	Morior	Teita	Tolai	Zulu	S.Jap.	
	3.1	3.7	5.4	5.5	5.9	5.9	6.2	6.2	
	0.77	0.43	0.01	0.01	0.00	0.00	0.00	0.00	
2688 Society Is.	Hawaii	Morior	Easter	Arikar	Guam	N.Jap.	S.Jap.	Anyang	
	3.7	4.2	5.6	6.3	6.5	6.8	6.8	6.9	
	0.44	0.20	0.00	0.00	-	-	-	-	
2689 Society Is.	Hawaii	Morior	Easter	N.Jap.	Guam	Ainu	S.Jap.	Tolai	
	4.9	5.5	6.1	6.6	6.7	6.9	7.4	7.5	
	0.04	0.00	0.00	-	-	-	-	-	
2690 Tuamotus	Easter	Hawaii	Anyang	Hainan	N.Jap.	Guam	S.Jap.	Morior	
	4.5	4.6	5.2	5.6	5.6	5.7	5.7	5.9	
	0.11	0.10	0.02	0.00	0.00	0.00	0.00	0.00	
2714 Hervey Is.	Arikar	Hainan	Anyang	Phil	Atayal	N.Jap.	Peru	S.Cruz	
	2.9	3.8	4.3	4.3	4.4	4.7	4.8	4.9	
	0.82	0.40	0.17	0.17	0.13	0.08	0.06	0.04	
2715 Marquesas	Hawaii	Easter	Morior	S.Jap.	Guam	Zulu	Hainan	Phil	
	4.4	4.7	5.0	5.2	5.4	5.4	5.5	5.5	
	0.15	0.06	0.03	0.02	0.01	0.01	0.01	0.00	
1650 Marquesas	Hawaii	Tolai	Morior	Tasman	Phil	Arikar	Guam	Peru	
	5.4	5.7	5.9	5.9	6.5	6.8	6.9	6.9	
	0.01	0.00	0.00	0.00	-	-	-	-	
2691 Marquesas	Morior	Hawaii	Guam	Ainu	Easter	Arikar	Peru	Hainan
	5.0	5.4	7.2	7.2	7.5	7.5	7.7	7.9
	0.03	0.00	-	-	-	-	-	-

TABLE 4-6

DISPOP results: Micronesia

2682 Guam	Anyang	Guam	Hainan	S.Jap.	Hawaii	Phil	N.Jap.	Atayal	
	4.1	4.3	4.6	4.9	5.2	5.3	5.3	5.9	
	0.27	0.18	0.10	0.05	0.02	0.01	0.01	0.00	
2683 Tinian	Guam	Ainu	Atayal	Phil	Hainan	Austr	S.Cruz	S.Jap.	
	3.9	4.8	5.0	5.1	5.3	5.3	5.3	5.4	
	0.34	0.06	0.03	0.02	0.01	0.01	0.01	0.01	
2684 Tinian	Guam	Phil	Anyang	Hainan	Hawaii	Zalvar	S.Jap.	Ainu	
	3.2	4.5	4.6	4.8	5.0	5.0	5.2	5.4	
	0.71	0.12	0.08	0.05	0.04	0.03	0.01	0.01	
2900 Marianas Saipan	Guam	Anyang	Easter	Hawaii	Hainan	S.Jap.	Atayal	N.Jap.	
	5.2	5.4	5.7	6.3	6.3	6.5	6.5	6.8	
	0.02	0.01	0.00	0.00	0.00	-	-	-	
2899 Marianas	Easter	Zalvar	Tolai	Guam	Ainu	Egypt	Norse	Zulu	
	5.1	5.6	5.9	6.1	6.2	6.2	6.2	6.3	
	0.03	0.00	0.00	0.00	0.00	0.00	0.00	0.00	
3325 Rota		Ainu	Guam	Hainan	S.Jap.	N.Jap.	Atayal	Egypt	Zulu
		3.9	4.1	4.8	4.9	5.4	5.4	5.7	5.8
		0.36	0.23	0.05	0.03	0.01	0.00	0.00	0.00
2901 Caroline Is.	Phil	Andamn	Hainan	S.Jap.	Anyang	Guam	Egypt	N.Jap.	
	4.3	4.7	4.8	4.9	5.0	5.0	5.0	5.4	
	0.18	0.07	0.05	0.04	0.03	0.03	0.03	0.01	
2902 Caroline Is.	Guam	Easter	Hawaii	Anyang	Hainan	S.Jap.	Phil	Tolai	
	3.5	3.6	4.0	4.7	4.8	4.9	5.1	5.3	
	0.56	0.50	0.33	0.07	0.06	0.05	0.02	0.01	
2051 Caroline Is.	Andamn	Phil	Egypt	Guam	Hawaii	Hainan	Zalvar	Easter	
	3.8	4.3	4.3	4.5	4.8	4.9	5.0	5.1	
	0.38	0.19	0.16	0.12	0.05	0.04	0.03	0.02	
2052 Caroline Is.	Teita	Andamn	Egypt	Morior	Norse	Guam	Eskimo	Easter	
	4.9	5.3	6.0	6.0	6.1	6.4	6.6	6.7	
	0.05	0.01	0.00	0.00	0.00	-	-	-	
2053 Gilbert Is.	Tolai	Easter	Guam	Hawaii	Anyang	S.Jap.	Hainan	Phil	
	4.6	5.0	5.2	5.3	5.9	6.0	6.1	6.3	
	0.09	0.03	0.02	0.01	0.00	0.00	0.00	0.00	
2054 Gilbert Is.	Andamn	Guam	Phil	Atayal	Hainan	Ainu	Peru	S.Jap.	
	4.1	4.1	4.3	4.4	4.4	4.5	4.5	4.5	
	0.26	0.25	0.17	0.16	0.15	0.12	0.12	0.12	
2713 Gilbert Is.	Egypt	Norse	Teita	Zalvar	S.Cruz	Peru	Andamn	Morior	
	3.9	4.2	5.0	5.2	5.6	5.7	5.7	5.8	
	0.36	0.21	0.03	0.02	0.00	0.00	0.00	0.00	
3191 Gilbert Is.	Guam	S.Jap.	Anyang	Eskimo	Andamn	Hainan	Phil	N.Jap.	
	4.6	5.1	5.3	5.5	5.5	5.6	5.6	5.7	
	0.10	0.03	0.01	0.01	0.00	0.00	0.00	0.00	
3051 Gilbert Is.		Hawaii	Easter	Morior	Tolai	Hainan	S.Jap.	Andamn	Eskimo
		4.6	5.1	5.5	5.6	5.9	5.9	6.0	6.1
		0.07	0.02	0.00	0.00	0.00	0.00	0.00	0.00

TABLE 4-7

DISPOP results: Boreal Peoples

1678 Eskimo Siberia	Eskimo 4.9 0.04	Arikar 5.1 0.02	N.Jap. 5.3 0.01	Buriat 5.6 0.00	S.Cruz 5.9 0.00	S.Jap. 5.9 0.00	Guam 5.9 0.00	Zalvar 6.0 0.00	
1679 Eskimo Siberia	Eskimo 6.4 -	Buriat 6.5 -	N.Jap. 7.8 -	S.Jap. 8.2 -	Norse 8.3 -	Tolai 8.4 -	Zalvar 8.5 -	Guam 8.5 -	
1682 Eskimo Pt. Barrow	Eskimo 4.3 0.19	Guam 6.3 0.00	Ainu 6.6 -	N.Jap. 6.6 -	Anyang 6.7 -	S.Jap. 6.7 -	Buriat 6.8 -	Arikar 7.0 -	
1680 Eskimo Siberia	Eskimo 2.8 0.88	Ainu 5.2 0.01	Arikar 5.4 0.01	Norse 5.5 0.00	N.Jap. 5.6 0.00	Zalvar 5.6 0.00	S.Jap. 5.6 0.00	Atayal 5.9 0.00	
1681 Eskimo Greenland	Eskimo 4.4 0.13	N.Jap. 6.2 -	Ainu 6.3 -	Arikar 6.4 -	S.Jap. 6.6 -	Norse 6.7 -	Atayal 6.7 -	Zalvar 6.9 -	
1684 Eskimo NW Greenland	Eskimo 3.1 0.79	Atayal 6.2 0.00	S.Jap. 6.2 0.00	Easter 6.2 0.00	N.Jap. 6.2 -	Ainu 6.5 -	Tolai 6.6 -	Hainan 6.8 -	
1685 Eskimo SW Greenland	Eskimo 3.8 0.43	Buriat 5.0 0.02	Arikar 5.3 0.01	Guam 5.9 0.00	Hainan 6.1 0.00	S.Jap. 6.1 0.00	Norse 6.4 -	N.Jap. 6.5 -	
2887 Ctrl Arctic	Eskimo 4.3 0.17	S.Cruz 4.5 0.13	S.Jap. 5.0 0.03	N.Jap. 5.2 0.02	Guam 5.3 0.01	Andamn 5.4 0.01	Atayal 5.4 0.01	Zalvar 5.5 0.01	
2888 Ctrl Arctic	Eskimo 5.4 0.01	N.Jap. 5.5 0.01	S.Jap. 5.6 0.00	S.Cruz 6.0 0.00	Atayal 6.3 0.00	Hainan 6.3 0.00	Buriat 6.4 -	Arikar 6.4 -	
2889 Ctrl Arctic	Eskimo 3.7 0.45	S.Jap. 6.6 -	Arikar 6.7 -	N.Jap. 6.7 -	Tolai 6.7 -	Easter 6.8 -	Atayal 6.8 -	S.Cruz 6.8 -	
2890 Ctrl Arctic	Eskimo 4.6 0.08	Ainu 4.8 0.05	Norse 5.0 0.03	S.Cruz 5.0 0.02	Zalvar 5.2 0.01	N.Jap. 5.5 0.00	S.Jap. 5.6 0.00	Arikar 5.7 0.00	
1683 Aleut	Peru 4.1 0.27	S.Cruz 4.2 0.22	Arikar 4.5 0.12	Hainan 4.8 0.05	S.Jap. 5.0 0.03	N.Jap. 5.2 0.02	Atayal 5.2 0.02	Phil 5.2 0.01	
1667 Chukchi	Eskimo 3.7 0.48	Guam 4.9 0.04	S.Jap. 5.2 0.01	N.Jap. 5.3 0.01	Arikar 5.3 0.01	S.Cruz 5.6 0.00	Ainu 5.7 0.00	Anyang 5.8 0.00	
1668 Chukchi	Eskimo 4.9 0.04	Norse 5.3 0.01	Zalvar 5.4 0.01	Guam 5.5 0.01	Arikar 5.6 0.00	Teita 5.8 0.00	Ainu 5.8 0.00	S.Cruz 5.9 0.00	
1669 Chukchi	Eskimo 4.9 0.05	Zalvar 6.3 0.00	Guam 6.5 -	Norse 6.5 -	Arikar 6.7 -	Egypt 7.1 -	Buriat 7.1 -	S.Cruz 7.3 -	
1670 Chukchi	Hainan 4.7 0.06	Guam 4.8 0.05	Arikar 4.9 0.04	Buriat 4.9 0.03	S.Jap. 5.3 0.01	N.Jap. 5.3 0.01	Zalvar 5.4 0.00	Hawaii 5.5 0.00	
1671 Chukchi	Buriat 5.7 0.00	Arikar 6.0 0.00	Eskimo 6.4 -	Hainan 6.6 -	S.Jap. 6.7 -	S.Cruz 6.7 -	N.Jap. 6.8 -	Guam 6.9 -	
1672 Chukchi	Eskimo 3.8 0.41	Ainu 4.9 0.03	Arikar 5.2 0.01	Zalvar 5.5 0.00	Norse 5.5 0.00	S.Cruz 5.7 0.00	S.Jap. 5.9 0.00	N.Jap. 6.0 0.00	

North Asia

Yakuts, Turkic-speakers of the upper Lena River, have no obviously expectable relations other than Buriats, with whom they were apparently formerly in contact. In fact they have almost no acceptable associations: one with Buriats and one with North Japanese are not inappropriate.

The Tungus-related Orochi and Orok of the shore opposite Japan and on Sakhalin, and the Moyoro shell heap remains, are all maritime people. Their probability values are low, but some nearest relations with Buriats and Japan is not inappropriate, if not diagnostic. At best, in a case of "unknowns," this would warn us that non-Asiatic assignment could be ruled out.

Japan

Problems of Ainu/Jomon/Japanese relations have been dealt with frequently (Howells 1966, 1986; Brace and Hunt 1990) with the most favored conclusion seeing modern Japanese as essentially being immigrants around the time of the appearance of the Yayoi culture (ca. 300 B.C. to A.D. 300). Here, an array of Jomon skulls, of different periods and localities, tends to affiliation with Ainu (seldom with high probability), but certainly not with Japanese. (The Uki specimen should probably be disregarded. It was in poor condition and was measured only because of its unusual, rather Hawaiian-like, appearance.) That is to say, the affiliation of choice is Ainu in just over half of the cases, if seldom strongly.

The question is not actual identification of Jomon skulls as Ainu. In temporal terms, all Ainu are Jomon, but not all Jomon are Ainu. There was doubtless a good deal of tribal variation among Jomon period peoples. The real point is the essential distinction of Jomon period skulls from later Japanese.

The picture changes with 20 Yayoi-culture crania. Two register as Ainu—one strongly—and almost all others connect with Japanese or Chinese, with good levels of probability. Secondary probabilities of Ainu connection are almost absent. (Number *2707*, the "Ainu," might in fact be a Jomon survivor, but a "Japanese" assignment is close behind in any case.) Cranial and archaeological evidence points strongly to a population replacement at this point in time.

One skull from the later Kofun period (ca. A.D. 300 to A.D. 700) is "Japanese," as expected. Three Koreans are Japanese/Chinese. An unidentified skull from Hakata Bay, site of the thirteenth-century "Kamikaze" ("divine wind" or typhoon) destruction of the invading Mongol fleet, seems clearly Japanese, not Mongol (which might call for a Buriat assignment.)

Crania from the Ryukyus (Kikai and Yoron Islands) are not solidly Japanese but seem more southern in affiliation—Anyang, Hainan, Guam—although almost all are well within a probability perimeter of Japanese assignment. Little information here, but a question of actual affinities of Ryukyu peoples would be worthy of study in any case. Present opinion places them closer to earlier peoples than are the main island Japanese.

South Chinese

The "Chinese" of this list (all male) were South Chinese laborers who died in America. The Folklo of Taiwan were of Fukienese derivation. All identify with the samples of China and of Atayal/Philippines origin. These populations in turn are not mutually well-differentiated. See figure 3 and table 4-10, below.

The skulls labeled "Anyang II–V" were long ago selected as type specimens by Li Chi from the mass of the Anyang collection, as representing hypothetical foreign ethnic contributions to the pool of sacrificial victims (and the population?) in the Shang tombs at Anyang (see Howells 1983). The perceived "types" were designated, respectively, Melanesoid, North European, Eskimoid, and undefined, conceivably Hawaiian-like. Metrical analysis does not support the visual assessment. All four type skulls have a good to high probability of being normal members of the Anyang population as a whole. (The Hainan Island Chinese population is close to Anyang in all analyses, as are both to Philippine Islanders.) Here we have a special case of hypothesis-testing by the present method.

TABLE 4-8

DISPOP results: North Asia

1661 Yakut	Eskimo	Norse	Buriat	Arikar	S.Cruz	Zalvar	Morior	Berg	
	6.6	6.7	6.7	7.1	7.1	7.2	7.4	7.4	
	-	-	-	-	-	-	-	-	
1662 Yakut	N.Jap.	S.Jap.	Guam	Buriat	Zalvar	Hainan	Phil	Norse	
	4.6	4.8	5.0	5.1	5.4	5.4	5.4	5.6	
	0.10	0.06	0.03	0.02	0.01	0.01	0.01	0.00	
1663 Yakut	Buriat	Eskimo	N.Jap.	S.Jap.	Guam	Norse	Ainu	Anyang	
	5.2	6.5	6.7	7.0	7.5	7.6	7.6	7.6	
	0.02	-	-	-	-	-	-	-	
1664 Yakut	Buriat	Guam	Arikar	Hainan	S.Jap.	N.Jap.	Eskimo	Ainu	
	3.4	6.6	6.6	6.6	6.8	6.9	7.2	7.6	
	0.62	-	-	-	-	-	-	-	
1665 Yakut	N.Jap.	Guam	Ainu	S.Jap.	Hainan	Arikar	Buriat	Morior	
	5.3	5.4	5.4	5.5	5.6	5.8	6.0	6.1	
	0.01	0.01	0.01	0.00	0.00	0.00	0.00	0.00	
1666 Yakut	Buriat	Eskimo	Arikar	N.Jap.	Guam	S.Jap.	Hainan	Morior	
	6.7	8.3	8.4	8.5	8.6	8.8	8.8	9.0	
	-	-	-	-	-	-	-	-	
1673 Orochi	Buriat	Eskimo	N.Jap.	S.Cruz	S.Jap.	Berg	Guam	Hainan	
	5.3	8.4	9.2	9.7	9.7	9.7	9.7	9.9	
	0.01	-	-	-	-	-	-	-	
1674 Orochi	Phil	Hainan	Arikar	Peru	Andamn	Anyang	S.Jap.	N.Jap.	
	4.1	4.4	4.9	5.1	5.2	5.2	5.2	5.3	
	0.25	0.15	0.04	0.02	0.01	0.01	0.01	0.01	
1675 Orochi	Buriat	Hainan	Guam	Arikar	S.Jap.	N.Jap.	Atayal	Ainu	
	4.4	4.9	5.1	5.3	5.9	6.0	6.1	6.2	
	0.13	0.03	0.02	0.01	0.00	0.00	0.00	0.00	
1676 Orochi	Buriat	Guam	Arikar	Hainan	Eskimo	S.Jap.	N.Jap.	Ainu	
	4.6	5.4	5.6	5.6	6.0	6.2	6.4	6.6	
	0.09	0.01	0.00	0.00	0.00	0.00	-	-	
1677 Orochi	Buriat	Hainan	N.Jap.	S.Jap.	Atayal	Eskimo	Guam	Arikar	
	5.5	5.8	6.2	6.2	6.3	6.3	6.3	6.7	
	0.00	0.00	0.00	-	-	-	-	-	
2056 Orok	Buriat	Arikar	Guam	Berg	Hainan	Anyang	N.Jap.	Phil	
	3.9	6.2	6.7	7.0	7.1	7.2	7.3	7.3	
	0.35	0.00	-	-	-	-	-	-	
2057 Moyoro I.	N.Jap.	S.Jap.	Eskimo	Guam	Hainan	Anyang	S.Cruz	Morior	
	5.6	5.7	5.9	6.2	6.3	6.4	6.5	6.5	
	0.00	0.00	0.00	0.00	0.00	-	-	-	
2058 Moyoro I.	S.Jap.	N.Jap.	Guam	Eskimo	Morior	Anyang	Hawaii	Hainan	
	5.9	5.9	6.0	6.1	6.1	6.2	6.7	6.7	
	0.00	0.00	0.00	0.00	0.00	0.00	-	-	
2059 Moyoro I.	N.Jap.	S.Jap.	Morior	Anyang	Guam	Hainan	Ainu	Phil	
	5.1	5.7	5.8	5.9	5.9	6.1	6.4	6.5	
	0.02	0.00	0.00	0.00	0.00	0.00	-	-	
2060 Moyoro I.	Buriat	N.Jap.	Eskimo	S.Jap.	Guam	Anyang	Hainan	Phil	
	6.4	6.6	6.8	6.8	7.1	7.1	7.3	7.9	
	-	-	-	-	-	-	-	-	
2061 Moyoro I.	Guam	Hainan	S.Jap.	N.Jap.	Arikar	Ainu	Atayal	S.Cruz	
	4.9	5.1	5.3	5.7	5.7	5.8	6.3	6.5	
	0.04	0.02	0.01	0.00	0.00	0.00	-	-	

TABLE 4-9

DISPOP results: Japan

2065 Tsukumo 3

Ainu	Norse	Egypt	Zulu	Zalvar	N.Jap.	Dogon	Morior
4.0	4.3	4.5	4.9	5.2	5.6	5.6	5.6
0.29	0.18	0.12	0.04	0.02	0.00	0.00	0.00

2066 Tsukumo 27

Ainu	Peru	S.Cruz	Berg	Atayal	Zalvar	Norse	Arikar
4.8	5.5	5.8	5.8	6.1	6.4	6.4	6.6
0.05	0.00	0.00	0.00	0.00	-	-	-

2067 Tsukumo 33

Norse	Ainu	Zalvar	Peru	Egypt	S.Cruz	Berg	Guam
4.2	4.8	5.0	5.2	5.2	5.8	6.1	6.2
0.20	0.06	0.03	0.02	0.02	0.00	0.00	0.00

2068 Tsukumo 58

Morior	Ainu	Peru	N.Jap.	S.Cruz	Anyang	Norse	Hainan
5.0	5.1	6.1	6.2	6.6	6.7	6.7	6.8
0.03	0.03	0.00	0.00	-	-	-	-

2069 Tsukumo 4

Arikar	Berg	Zalvar	Norse	Peru	Ainu	N.Jap.	S.Jap.
4.7	5.2	5.3	5.4	5.7	5.9	6.2	6.2
0.07	0.01	0.01	0.00	0.00	0.00	0.00	-

2070 Tsukumo 14

Ainu	Arikar	Zalvar	Berg	Norse	Atayal	N.Jap.	S.Jap.
3.2	3.3	3.8	3.8	4.2	4.6	4.7	4.9
0.74	0.66	0.41	0.40	0.20	0.09	0.06	0.04

2071 Tsukumo 34

Arikar	Ainu	Atayal	S.Jap.	Berg	Norse	Teita	N.Jap.
4.2	4.2	5.2	5.2	5.2	5.3	5.4	5.6
0.23	0.19	0.01	0.01	0.01	0.01	0.01	0.00

2062 Takasago 1

Ainu	Norse	Morior	Zalvar	N.Jap.	Egypt	Berg	S.Jap.
4.9	6.8	7.0	7.1	7.2	7.2	7.3	7.5
0.04	-	-	-	-	-	-	-

2063 Rebunge

Ainu	Guam	Norse	Zalvar	Egypt	Zulu	Easter	Hawaii
4.9	6.1	6.2	6.4	6.5	6.6	6.6	6.8
0.04	0.00	0.00	0.00	-	-	-	-

2064 Onkoromanai

Norse	Ainu	Zalvar	Egypt	Berg	Peru	Austr	Zulu
4.7	4.7	5.3	5.7	6.1	6.8	6.9	7.1
0.08	0.07	0.01	0.00	0.00	-	-	-

2072 Yoshigo

Ainu	Peru	Phil	Hainan	N.Jap.	Norse	Egypt	S.Cruz
5.7	5.9	6.3	6.4	6.5	6.5	6.6	6.6
0.00	0.00	0.00	-	-	-	-	-

2073 Uki

Anyang	Guam	Hainan	S.Jap.	Phil	N.Jap.	Atayal	Arikar
3.5	3.6	3.9	4.4	4.8	4.8	5.0	5.3
0.54	0.48	0.36	0.14	0.06	0.05	0.03	0.01

2074 Ota Jomon

Ainu	Morior	Arikar	S.Jap.	N.Jap.	Atayal	Peru	Zalvar
4.9	5.0	5.4	5.7	5.8	5.8	5.8	5.8
0.04	0.02	0.01	0.00	0.00	0.00	0.00	0.00

2692 Yayoi

Phil	N.Jap.	Hainan	S.Jap.	Atayal	Anyang	S.Cruz	Peru
3.6	3.8	4.1	4.2	4.5	4.9	5.1	5.3
0.53	0.40	0.27	0.21	0.11	0.04	0.02	0.01

2693 Yayoi

N.Jap.	S.Jap.	Phil	Anyang	Hainan	Guam	Atayal	Ainu
3.7	3.8	3.9	4.2	4.2	4.4	5.0	5.0
0.48	0.39	0.37	0.23	0.22	0.14	0.03	0.03

2694 Yayoi

N.Jap.	Anyang	S.Jap.	Hainan	Ainu	Atayal	Guam	Phil
3.9	4.4	4.7	4.8	5.0	5.0	5.5	5.7
0.33	0.14	0.07	0.06	0.03	0.03	0.00	0.00

2695 Yayoi

Anyang	S.Jap.	Hainan	Atayal	N.Jap.	Phil	Ainu	Guam
4.3	5.0	5.1	5.2	5.7	6.1	6.2	6.5
0.16	0.03	0.02	0.02	0.00	0.00	0.00	-

2696 Yayoi

Guam	N.Jap.	Hainan	Phil	S.Jap.	Anyang	Atayal	Arikar
3.7	3.8	4.0	4.0	4.2	4.2	4.8	5.0
0.43	0.40	0.32	0.28	0.22	0.21	0.05	0.03

2697 Yayoi

Ainu	Hainan	S.Jap.	Anyang	Atayal	Guam	N.Jap.	Dogon
4.2	4.5	4.6	4.6	4.8	4.8	4.8	5.0
0.21	0.11	0.10	0.10	0.06	0.06	0.06	0.03

TABLE 4-9 CONTINUED
DISPOP results: Japan

2698 Yayoi	Phil 5.1 0.02	Buriat 5.6 0.00	Arikar 5.6 0.00	Hainan 5.8 0.00	N.Jap. 6.1 0.00	S.Jap. 6.3 0.00	Anyang 6.4 -	Andamn 6.4 -	
2699 Yayoi	Buriat 4.3 0.18	N.Jap. 4.5 0.12	S.Jap. 4.7 0.08	Phil 4.7 0.07	Hainan 4.8 0.06	Guam 5.2 0.01	Anyang 5.3 0.01	Atayal 5.5 0.01	
2700 Yayoi	N.Jap. 4.4 0.14	Phil 4.6 0.08	S.Jap. 4.8 0.06	S.Cruz 4.8 0.06	Guam 4.8 0.06	Buriat 4.9 0.05	Hainan 4.9 0.04	Peru 5.1 0.02	
2707 Yayoi	Ainu 2.4 0.94	S.Jap. 3.4 0.60	N.Jap. 3.5 0.58	Hainan 3.5 0.54	Phil 3.7 0.47	Atayal 3.7 0.46	Guam 3.7 0.44	Anyang 3.8 0.42	
2708 Yayoi	Phil 3.8 0.38	Zalvar 4.2 0.21	Ainu 4.3 0.17	Egypt 4.4 0.14	Norse 4.4 0.14	Guam 4.5 0.11	Zulu 4.6 0.10	Dogon 4.6 0.09	
2709 Yayoi	N.Jap. 5.2 0.02	S.Jap. 5.4 0.01	Zulu 5.7 0.00	Guam 5.8 0.00	Ainu 6.0 0.00	Hainan 6.3 0.00	Anyang 6.4 -	Atayal 6.5 -	
2710 Yayoi	Anyang 4.5 0.13	Hainan 4.5 0.11	Atayal 4.7 0.07	Phil 4.8 0.06	S.Jap. 5.0 0.03	N.Jap. 5.4 0.01	Guam 5.9 0.00	Buriat 6.3 0.00	
2711 Yayoi	N.Jap. 4.1 0.27	S.Jap. 4.1 0.26	Hainan 4.4 0.13	Anyang 4.4 0.13	Phil 4.6 0.09	Guam 5.6 0.00	Atayal 5.7 0.00	Dogon 6.3 0.00	
2701 Yayoi	S.Jap. 4.2 0.19	Arikar 4.3 0.17	Atayal 4.9 0.04	Hainan 4.9 0.03	N.Jap. 5.0 0.03	Guam 5.2 0.01	Ainu 5.2 0.01	Morior 5.2 0.01	
2702 Yayoi	S.Jap. 3.5 0.55	Hainan 3.8 0.43	N.Jap. 3.8 0.39	Guam 3.9 0.35	Arikar 4.1 0.25	Ainu 4.8 0.05	Zulu 4.8 0.04	Atayal 4.9 0.03	
2703 Yayoi	Hainan 4.8 0.05	Arikar 4.8 0.05	Peru 5.3 0.01	Buriat 5.7 0.00	S.Jap. 5.7 0.00	Atayal 6.0 0.00	Guam 6.0 0.00	Berg 6.4 -	
2704 Yayoi	S.Jap. 4.0 0.30	Arikar 4.1 0.27	Guam 4.1 0.25	N.Jap. 4.1 0.24	Hainan 4.1 0.24	Morior 4.7 0.06	Ainu 4.8 0.04	Peru 5.2 0.01	
2705 Yayoi	S.Jap. 3.8 0.41	N.Jap. 4.2 0.23	Guam 4.8 0.05	Hainan 5.0 0.03	Atayal 5.6 0.00	Ainu 5.7 0.00	Zulu 5.9 0.00	Norse 6.4 -	
2706 Yayoi	Hainan 4.5 0.11	S.Jap. 4.5 0.10	Dogon 5.0 0.02	N.Jap. 5.0 0.02	Atayal 5.1 0.02	Guam 5.1 0.01	Zulu 5.5 0.00	Ainu 5.7 0.00	
2075 Kofun Period	S.Jap. 4.0 0.30	N.Jap. 4.4 0.15	Guam 5.2 0.01	Hainan 5.4 0.01	Anyang 5.5 0.01	Phil 5.5 0.00	Atayal 6.3 0.00	Hawaii 6.6 -	
2628 Korea	S.Jap. 4.1 0.27	N.Jap. 4.2 0.20	Hainan 4.5 0.13	Anyang 4.8 0.05	Atayal 4.9 0.04	Phil 4.9 0.04	Guam 5.2 0.02	Arikar 5.7 0.00	
2629 Korea	Hainan 4.0 0.32	S.Jap. 4.0 0.29	Anyang 4.3 0.16	Guam 4.4 0.16	Phil 4.5 0.11	Atayal 4.6 0.08	N.Jap. 5.1 0.02	Zulu 6.3 0.00	
2630 Korea	N.Jap. 3.2 0.73	Anyang 3.4 0.60	S.Jap. 3.5 0.54	Hainan 3.9 0.35	Phil 4.5 0.13	Atayal 4.5 0.11	Easter 5.3 0.01	Guam 5.4 0.01	
2712 Unknown (Hakata Bay)	S.Jap. 3.2 0.69	Anyang 3.5 0.57	N.Jap. 3.8 0.40	Hainan 4.1 0.28	Guam 4.2 0.22	Phil 4.4 0.14	Atayal 5.0 0.03	Egypt 5.1 0.02	

TABLE 4-9 CONTINUED
DISPOP results: Japan

2633 Ryukyus Kikai	Guam 4.6 0.08	Ainu 5.0 0.03	Easter 5.0 0.03	Anyang 5.1 0.02	Atayal 5.6 0.00	Hainan 5.6 0.00	Phil 5.7 0.00	Zalvar 5.9 0.00	
2634 Ryukyus Kikai	Anyang 3.0 0.81	N.Jap. 3.0 0.78	Hainan 3.1 0.76	S.Jap. 3.1 0.76	Atayal 3.2 0.71	Phil 3.7 0.46	Guam 4.1 0.26	Zalvar 4.6 0.09	
2635 Ryukyus Kikai	Teita 3.7 0.47	S.Jap. 4.0 0.32	Anyang 4.0 0.31	Guam 4.1 0.27	Hainan 4.1 0.25	Atayal 4.2 0.22	Zulu 4.2 0.20	Phil 4.3 0.17	
2636 Ryukyus Kikai	Hainan 2.2 0.97	S.Jap. 2.3 0.97	N.Jap. 2.5 0.93	Anyang 2.6 0.92	Atayal 3.0 0.79	Phil 3.1 0.77	Guam 3.8 0.43	Ainu 3.9 0.35	
2637 Ryukyus Kikai	N.Jap. 2.7 0.89	S.Jap. 2.8 0.87	Hainan 3.2 0.73	Phil 3.2 0.70	Anyang 3.6 0.51	Atayal 4.2 0.22	Guam 4.3 0.18	Andamn 4.9 0.04	
2639 Ryukyus Yoron I.	Anyang 2.6 0.91	Hainan 2.8 0.87	Phil 2.8 0.86	S.Jap. 3.4 0.61	N.Jap. 3.6 0.53	Guam 3.6 0.52	Atayal 3.7 0.47	Ainu 4.5 0.12	
2640 Ryukyus Yoron I.	Guam 4.9 0.04	N.Jap. 5.0 0.03	Phil 5.6 0.00	Hawaii 5.7 0.00	S.Jap. 5.7 0.00	Anyang 5.9 0.00	Hainan 6.2 0.00	Ainu 6.2 0.00	
2638 Ryukyus Kikai	Guam 3.4 0.60	Ainu 4.0 0.31	S.Jap. 4.2 0.21	Hainan 4.4 0.15	N.Jap. 4.5 0.11	Norse 5.7 0.00	Arikar 5.8 0.00	Zulu 5.8 0.00
2641 Ryukyus Yoron I.	Hainan 3.7 0.44	S.Jap. 3.8 0.42	Teita 4.1 0.27	Guam 4.1 0.26	Atayal 4.3 0.18	N.Jap. 4.7 0.07	Dogon 4.8 0.04	Andamn 5.0 0.02
2642 Ryukyus Yoron I.	S.Jap. 2.2 0.98	N.Jap. 2.9 0.85	Hainan 3.1 0.77	Atayal 3.4 0.61	Guam 3.5 0.57	Ainu 4.5 0.09	Zulu 4.9 0.04	Zalvar 5.0 0.03

TABLE 4-10
DISPOP results: China

1784 Chinese	Phil 3.9 0.33	Guam 4.0 0.30	S.Jap. 4.1 0.25	Hainan 4.1 0.25	Atayal 4.1 0.24	Anyang 4.5 0.12	N.Jap. 4.5 0.12	Zulu 5.4 0.01
1785 Chinese	Phil 3.3 0.65	Hainan 3.4 0.63	Berg 4.2 0.23	Atayal 4.2 0.21	Anyang 4.3 0.19	Arikar 4.3 0.17	Zalvar 4.3 0.16	S.Jap. 4.5 0.12
1786 Chinese	Anyang 3.9 0.36	S.Jap. 4.4 0.14	Hainan 4.5 0.11	N.Jap. 5.2 0.02	Phil 5.2 0.02	Atayal 5.4 0.01	Guam 6.1 0.00	Easter 6.5 -
1787 Chinese	Atayal 3.6 0.52	Hainan 4.0 0.32	Anyang 4.4 0.14	Phil 4.7 0.08	S.Jap. 5.3 0.01	N.Jap. 6.1 0.00	Arikar 6.1 0.00	Guam 6.2 0.00
1788 Chinese	S.Jap. 4.6 0.10	Anyang 4.6 0.10	Phil 4.7 0.08	Hainan 4.8 0.06	N.Jap. 4.8 0.05	Atayal 5.5 0.00	Zulu 6.0 0.00	Hawaii 6.1 0.00

TABLE 4-10 CONTINUED
DISPOP results: China

1789 Chinese	N.Jap.	S.Jap.	Hainan	Anyang	Phil	Atayal	Guam	S.Cruz
	2.4	2.5	2.9	3.0	3.3	3.4	3.9	4.7
	0.95	0.93	0.84	0.79	0.68	0.63	0.36	0.07
2643 Folklo	Anyang	Hainan	S.Jap.	Phil	Guam	Atayal	N.Jap.	Peru
	2.3	2.6	3.5	3.6	3.9	4.0	4.1	5.0
	0.97	0.92	0.56	0.49	0.35	0.32	0.26	0.03
2644 Folklo	Phil	Hainan	Anyang	S.Jap.	Atayal	Guam	N.Jap.	Zulu
	3.3	3.5	3.6	3.8	4.3	4.4	4.4	4.6
	0.69	0.55	0.50	0.41	0.19	0.15	0.15	0.10
2645 Folklo	Hainan	Anyang	Phil	S.Jap.	Atayal	Guam	N.Jap.	Arikar
	2.2	2.7	3.0	3.1	3.7	3.7	4.1	5.0
	0.97	0.90	0.81	0.74	0.46	0.45	0.26	0.03
2646 Folklo	Phil	Hainan	Anyang	Guam	Atayal	S.Jap.	Andamn	Arikar
	4.1	4.3	4.5	4.7	5.4	5.6	6.2	6.6
	0.25	0.18	0.10	0.07	0.01	0.00	0.00	-
2647 Folklo	Atayal	Hainan	Anyang	Andamn	S.Jap.	Phil	N.Jap.	Teita
	3.4	3.4	3.7	3.7	3.8	4.2	4.9	5.1
	0.64	0.63	0.47	0.45	0.42	0.23	0.05	0.02
2648 Folklo	Hainan	Atayal	S.Jap.	Anyang	Phil	N.Jap.	Arikar	Zalvar
	3.8	4.3	4.5	4.6	5.0	5.0	5.4	5.5
	0.39	0.20	0.12	0.08	0.03	0.03	0.01	0.00
2650 Anyang II	Phil	Hainan	S.Jap.	Atayal	Anyang	Dogon	Zulu	Andamn
	2.9	3.2	3.6	4.0	4.0	4.1	4.3	4.5
	0.85	0.71	0.50	0.32	0.31	0.24	0.16	0.12
2651 Anyang III	Hainan	Anyang	Atayal	S.Jap.	Phil	Peru	Guam	Andamn
	2.8	3.4	3.4	3.6	3.7	4.1	4.3	4.4
	0.88	0.61	0.60	0.51	0.43	0.28	0.18	0.16
2652 Anyang IV	Anyang	Hainan	S.Jap.	Atayal	N.Jap.	Guam	Phil	Morior
	3.4	4.3	4.5	4.6	4.8	5.2	5.4	5.5
	0.62	0.18	0.13	0.08	0.06	0.01	0.01	0.01
2653 Anyang V	Anyang	Hainan	Phil	Atayal	S.Jap.	Guam	N.Jap.	Arikar
	3.4	3.7	4.1	4.2	4.3	4.4	4.8	5.2
	0.61	0.46	0.25	0.20	0.17	0.13	0.05	0.01

Taiwan

This is a list of aboriginals of Taiwan, partly identified by tribe in the collections, and partly not so recorded. They are almost uniformly specified by DISPOP as Atayal, Philippines, or Hainan Chinese in first and second choice. The general closeness of all these populations, seen already in other ways, is thus confirmed here, and so specification more closely as one or another of them is not usually to be expected. Only a Bunun skull, out of the 19 shown, diverges significantly, with an allocation as Bush. Bunun are rather exceptional among Taiwan aboriginals, as to small size and dark skin color. (See Chai 1967.)

The Banton Island skull is Filipino; for affiliation, Atayal is appropriate enough, but Philippine skulls are the most nondescript among our samples.

America

These results are more puzzling. They seem to suggest, as do other figures, that there is wide cranial variability in the people of the hemisphere that is not comprehended by the variation among or within the three samples available for this study. The Indian Knoll specimens of this early population align themselves satisfactorily with American populations in two cases, and with Mongoloids more generally in the third. A female

TABLE 4-11

DISPOP results: Taiwan

2665 Bunun	Atayal	S.Jap.	Hainan	N.Jap.	Anyang	Phil	Ainu	Guam	
	2.8	3.0	3.3	3.5	3.7	3.8	3.8	4.0	
	0.88	0.81	0.68	0.58	0.45	0.43	0.40	0.32	

2666 Bunun	Bush	Ainu	Atayal	Phil	Zalvar	S.Cruz	Norse	Hainan	
	4.4	4.7	4.8	4.9	5.2	5.2	5.3	5.4	
	0.15	0.07	0.05	0.04	0.02	0.02	0.01	0.01	

2667 Bunun	N.Jap.	S.Jap.	Atayal	Hainan	Ainu	Andamn	Guam	Arikar
	3.1	3.2	3.4	3.5	4.3	4.7	4.8	4.9
	0.79	0.75	0.63	0.60	0.18	0.06	0.04	0.04

2668 Paiwan	Phil	Atayal	Hainan	S.Jap.	Anyang	N.Jap.	Guam	Zulu	
	2.9	3.0	3.3	3.4	3.7	4.1	4.1	4.2	
	0.82	0.81	0.67	0.59	0.46	0.27	0.26	0.20	

2669 Yami	Atayal	Hainan	Phil	Guam	Anyang	S.Jap.	Andamn	N.Jap.	
	3.8	3.9	3.9	4.1	4.2	4.4	4.8	5.1	
	0.38	0.35	0.34	0.25	0.23	0.13	0.05	0.02	

2670 Yami	Phil	Hainan	Andamn	Atayal	Anyang	S.Jap.	Guam	N.Jap.	
	3.6	3.6	4.1	4.3	4.4	4.5	5.5	5.5	
	0.54	0.49	0.24	0.18	0.14	0.12	0.00	0.00	

2671 Yami	Phil	Hainan	Andamn	Anyang	Atayal	Guam	S.Jap.	N.Jap.	
	2.6	3.2	3.7	4.0	4.1	4.1	4.2	4.6	
	0.92	0.70	0.48	0.33	0.26	0.25	0.22	0.10	

2672 Yami	Hainan	Guam	S.Jap.	Atayal	N.Jap.	Ainu	Andamn	Hawaii
	3.6	3.7	4.9	5.5	5.6	6.0	6.4	6.7
	0.50	0.46	0.04	0.00	0.00	0.00	-	-

2673 Yami	Atayal	Hainan	S.Jap.	Guam	N.Jap.	Arikar	Hawaii	Andamn
	3.3	3.7	4.2	4.6	4.9	5.1	5.1	5.4
	0.70	0.44	0.20	0.08	0.03	0.02	0.02	0.01

2655 Taiwan	Atayal	Anyang	Hainan	S.Jap.	Phil	N.Jap.	Andamn	Bush	
	3.4	3.9	4.1	4.6	4.8	5.4	5.8	5.9	
	0.64	0.37	0.24	0.09	0.05	0.01	0.00	0.00	

2656 Taiwan	Atayal	Hainan	Phil	Anyang	S.Jap.	N.Jap.	Zulu	Guam	
	2.4	2.5	2.7	3.2	3.3	4.4	4.5	4.5	
	0.94	0.94	0.90	0.70	0.66	0.14	0.13	0.10	

2657 Taiwan	Atayal	Zalvar	Peru	Hainan	Arikar	S.Cruz	Phil	Norse	
	3.9	4.0	4.1	4.4	4.6	4.6	4.7	4.8	
	0.36	0.28	0.26	0.14	0.10	0.09	0.07	0.05	

2658 Taiwan	Zalvar	Atayal	Phil	Hainan	Ainu	S.Jap.	N.Jap.	Zulu	
	3.6	3.7	3.7	3.8	3.9	4.0	4.3	4.3	
	0.53	0.47	0.45	0.38	0.36	0.33	0.19	0.19	

2659 Taiwan	Atayal	Phil	Hainan	S.Jap.	Zulu	Anyang	Peru	N.Jap.	
	3.2	3.7	3.8	4.0	4.1	4.2	4.3	4.3	
	0.71	0.46	0.38	0.29	0.26	0.21	0.19	0.19	

2660 Taiwan	Atayal	Phil	Andamn	Hainan	S.Cruz	Peru	S.Jap.	Anyang	
	4.2	4.3	4.6	4.7	4.9	5.2	5.3	5.4	
	0.22	0.18	0.09	0.07	0.05	0.02	0.01	0.01	

2661 Taiwan	Hainan	Phil	Guam	Atayal	Peru	Anyang	Teita	Andamn	
	5.6	5.6	5.7	6.0	6.3	6.4	6.4	6.6	
	0.00	0.00	0.00	0.00	0.00	0.00	-	-	

2662 Taiwan	Hainan	Atayal	Phil	S.Jap.	Anyang	Andamn	Zalvar	Egypt	
	3.0	3.0	3.2	3.6	3.7	3.8	4.2	4.4	
	0.80	0.80	0.71	0.49	0.48	0.39	0.21	0.14	

2663 Taiwan	Hainan	Guam	S.Jap.	Atayal	Ainu	N.Jap.	Dogon	Zulu
	3.8	3.9	4.1	4.3	4.4	4.7	5.4	5.5
	0.42	0.34	0.25	0.18	0.13	0.07	0.01	0.00

TABLE 4-11 CONTINUED
DISPOP results: Taiwan

2664 Taiwan	Atayal	Ainu	S.Jap.	Guam	Hainan	Zulu	N.Jap.	Zalvar
	3.4	4.3	4.4	4.4	4.6	4.8	5.1	5.8
	0.64	0.16	0.14	0.12	0.08	0.05	0.02	0.00
2674 Banton I. (Phil. Is.)	Atayal	S.Jap.	Tolai	Zulu	Arikar	N.Jap.	Zalvar	Ainu
	4.6	4.7	4.8	4.9	5.1	5.3	5.4	5.5
	0.08	0.07	0.05	0.03	0.02	0.01	0.00	0.00

TABLE 4-12
DISPOP results: America

1790 Indian Knoll	S.Cruz	Arikar	Peru	Zalvar	Norse	Eskimo	Andamn	Guam
	3.8	4.2	4.4	4.8	4.9	5.2	5.2	5.3
	0.39	0.21	0.16	0.05	0.04	0.02	0.02	0.01
1791 Indian Knoll	Hainan	S.Jap.	Atayal	Phil	N.Jap.	Anyang	Peru	Arikar
	2.9	3.0	3.1	3.6	3.6	3.7	3.9	4.3
	0.85	0.80	0.75	0.54	0.53	0.46	0.37	0.16
1792 Indian Knoll	Arikar	Peru	S.Cruz	Zalvar	Guam	Hainan	Norse	Phil
	3.8	4.2	4.8	4.8	4.9	5.0	5.0	5.3
	0.42	0.20	0.06	0.05	0.04	0.03	0.03	0.01
1793 Indian Maine	Norse	Zalvar	Morior	Arikar	Berg	Egypt	Ainu	Peru
	3.4	3.9	4.1	4.2	4.5	4.5	4.6	4.9
	0.62	0.35	0.26	0.20	0.11	0.11	0.08	0.04
1794 Indian Massachusetts	Tolai	Peru	Easter	Arikar	Morior	S.Cruz	Tasman	Guam
	4.2	4.9	5.5	5.7	5.7	5.8	5.8	5.9
	0.20	0.04	0.01	0.00	0.00	0.00	0.00	0.00
1865 Indian Illinois	Easter	S.Jap.	N.Jap.	Anyang	Hawaii	Hainan	Arikar	Guam
	4.8	4.9	5.2	5.4	5.5	5.5	5.6	5.6
	0.06	0.04	0.02	0.01	0.01	0.01	0.00	0.00
3326 Pecos	Anyang	S.Jap.	Hainan	Guam	Phil	N.Jap.	Atayal	Easter
	3.7	4.0	4.0	4.3	4.6	4.8	4.8	5.1
	0.46	0.33	0.32	0.18	0.08	0.06	0.05	0.02
3327 Pecos	Arikar	Peru	Hainan	Atayal	S.Jap.	Anyang	Guam	Phil
	4.4	4.7	4.9	5.1	5.6	5.6	5.8	5.8
	0.15	0.07	0.04	0.02	0.00	0.00	0.00	0.00
3328 Pecos	S.Jap.	Hainan	Atayal	N.Jap.	Anyang	Easter	Hawaii	Guam
	5.2	6.0	6.0	6.1	6.1	6.6	6.8	6.9
	0.02	0.00	0.00	0.00	0.00	-	-	-
3329 Pecos	Zalvar	Berg	Norse	Peru	Egypt	S.Cruz	Atayal	Hainan
	3.5	4.4	4.5	4.6	5.0	5.1	5.3	5.6
	0.55	0.16	0.11	0.08	0.03	0.02	0.01	0.00
3330 Pecos	Peru	Hainan	Anyang	N.Jap.	Atayal	S.Jap.	Phil	Arikar
	4.1	4.5	4.7	5.0	5.1	5.1	5.4	5.5
	0.26	0.11	0.07	0.03	0.02	0.02	0.01	0.00
3331 Pecos	Atayal	Hainan	N.Jap.	S.Jap.	Anyang	Ainu	Peru	Phil
	3.4	4.4	4.5	4.5	4.5	4.9	5.3	5.6
	0.59	0.14	0.11	0.11	0.11	0.04	0.01	0.00
3332 Pecos	Arikar	Peru	Hainan	Hawaii	S.Jap.	Guam	S.Cruz	N.Jap.
	4.8	5.1	6.0	6.1	6.2	6.2	6.3	6.4
	0.05	0.02	0.00	0.00	0.00	0.00	0.00	-
3333 Pecos	Guam	Arikar	Hainan	S.Jap.	Buriat	Anyang	Phil	Atayal
	4.7	4.8	4.9	5.1	5.1	5.1	5.2	5.4
	0.07	0.05	0.05	0.02	0.02	0.02	0.02	0.01

TABLE 4-13
DISPOP results: South and Southeast Asia

1615 Madras	Teita	Egypt	Zalvar	Norse	Andamn	Phil	Bush	Zulu	
	4.5	4.7	5.0	5.5	5.6	5.9	6.1	6.2	
	0.11	0.08	0.03	0.00	0.00	0.00	0.00	0.00	
1616 Madras	Zalvar	Egypt	Norse	Berg	Phil	Hainan	Teita	Guam	
	5.0	5.2	5.3	5.5	5.6	5.8	5.9	5.9	
	0.03	0.02	0.01	0.01	0.00	0.00	0.00	0.00	
1617 Madras		Zulu	Egypt	Bush	Norse	Zalvar	Dogon	Teita	Ainu
		4.0	4.0	4.4	4.7	5.0	5.0	5.3	5.4
		0.31	0.30	0.14	0.06	0.03	0.02	0.01	0.01
1618 Madras		Egypt	Norse	Zalvar	Ainu	Zulu	Easter	Atayal	Teita
		4.2	4.3	4.3	4.7	4.9	5.5	5.5	5.5
		0.22	0.19	0.18	0.07	0.03	0.00	0.00	0.00
1619 Vedda	Andamn	Phil	Egypt	Zalvar	Atayal	Dogon	Hainan	Zulu	
	3.5	4.6	4.8	4.9	5.0	5.4	5.4	5.6	
	0.57	0.09	0.06	0.04	0.03	0.01	0.01	0.00	
1620 Vedda	Egypt	Andamn	Dogon	Teita	Phil	Hainan	Ainu	Guam	
	4.1	4.1	4.9	4.9	5.3	5.4	5.4	5.5	
	0.25	0.24	0.04	0.04	0.01	0.01	0.01	0.01	
1621 Vedda		Ainu	Egypt	Atayal	Teita	Norse	Guam	Zalvar	Zulu
		4.9	5.2	5.5	5.5	5.6	5.6	5.7	5.8
		0.03	0.01	0.00	0.00	0.00	0.00	0.00	0.00
1622 Vedda		Egypt	Zalvar	Norse	Ainu	Atayal	Andamn	Bush	Dogon
		3.7	4.8	5.0	5.3	5.4	5.7	5.8	5.8
		0.48	0.04	0.03	0.01	0.01	0.00	0.00	0.00
1623 Burma	Peru	Arikar	Phil	Zalvar	Hainan	S.Cruz	Berg	Atayal	
	5.5	5.5	5.6	5.8	6.0	6.2	6.2	6.2	
	0.01	0.01	0.00	0.00	0.00	0.00	0.00	0.00	
1624 Burma	Egypt	Teita	Zalvar	Norse	Zulu	Austr	Phil	Dogon	
	4.3	4.4	4.7	4.9	4.9	5.0	5.2	5.4	
	0.17	0.16	0.07	0.05	0.05	0.03	0.02	0.01	
1625 Burma	Phil	Andamn	Guam	Egypt	Hainan	Zalvar	Arikar	Hawaii	
	5.0	5.2	5.4	5.7	5.8	5.8	6.0	6.0	
	0.03	0.02	0.01	0.00	0.00	0.00	0.00	0.00	
1626 Burma	Egypt	Zulu	Teita	Zalvar	Norse	S.Jap.	Dogon	Andamn	
	3.9	5.1	5.2	5.3	5.5	6.1	6.3	6.3	
	0.33	0.02	0.02	0.01	0.00	0.00	0.00	0.00	
1627 Burma		Berg	S.Cruz	Zalvar	Peru	Arikar	Norse	Atayal	Egypt
		3.7	4.3	4.4	4.8	4.8	4.8	5.0	5.0
		0.48	0.16	0.15	0.05	0.05	0.05	0.02	0.02
1628 Burma		Egypt	Andamn	Norse	Hainan	Zalvar	Ainu	Berg	Atayal
		4.0	4.8	4.9	5.1	5.1	5.2	5.3	5.4
		0.28	0.05	0.03	0.02	0.01	0.01	0.01	0.01
1629 Dyak (Borneo)	Arikar	Phil	Berg	Hainan	Atayal	Peru	S.Cruz	Zalvar	
	3.7	5.4	5.4	5.5	5.6	6.0	6.1	6.1	
	0.46	0.01	0.01	0.01	0.00	0.00	0.00	0.00	
1630 Dyak (Borneo)		Guam	S.Jap.	N.Jap.	Ainu	Hainan	Atayal	Norse	Zalvar
		4.7	4.8	4.9	5.3	5.4	5.5	5.5	5.6
		0.07	0.04	0.03	0.01	0.01	0.00	0.00	0.00
1631 Battak (Sumatra)	Zalvar	Norse	N.Jap.	Ainu	Guam	Egypt	Morior	S.Jap.	
	4.0	4.1	4.2	4.2	4.3	4.3	4.3	4.3	
	0.29	0.27	0.22	0.21	0.18	0.18	0.18	0.17	
1632 Atchinese (Sumatra)	Teita	Egypt	Zalvar	Guam	Zulu	Easter	Norse	Phil	
	4.6	4.8	5.0	5.2	5.3	5.4	5.5	5.5	
	0.09	0.06	0.03	0.02	0.01	0.01	0.01	0.00	

from Tierra del Fuego satisfactorily has all three American populations as nearest neighbors. The other single specimens make no sense at all.

The same holds for the Pecos Pueblo sample. Two things may be said. First, this population survived into the twentieth century, well into the period of contact, when Spanish or other Indian genes may have been incorporated. Second, the great majority of Pecos (and other Pueblo) crania were strongly deformed by cradling practices. The specimens used here were selected by looking in Hooton's records for "undeformed" cases and by trying to make sure that these were indeed unaffected. But there may well have been undetected effects, and Brace and Hunt (1990) also found Southwestern Indian skulls to give similarly undisciplined results. I consider the present readings to be essentially worthless. About all that can be said is that Polynesians, Australoids, and Africans are ruled out as connections, and if #3329 is actually a Spaniard (which he does not appear to be), then Indian "Europeans" also do not appear. That leaves Indians and Asiatics as assignments, which is not very close information.

South and Southeast Asia

Here is an area not represented among the base populations, beyond the expectation that South China or the Philippines represent that part of the world. In diagnosis, these cases seem neither expectable nor informative.

For four natives of Southern India (Madras) the possible affiliations are Egypt or sub-Saharan Africa. For four Veddas of Sri Lanka, assignment is Egypt or Andaman Islands, an interesting note, of which one can make what one wishes. However, Veddas give no indication at all of an "Australoid" connection, one sometimes suggested, specifically by myself (Howells 1937), so this notion can be put aside.

For six Burmese, the only probabilities greater than .03 are for "Europeans": Egypt or Berg. Farther to the east, if Filipinos or Atayals might have been expected to serve as bellwethers for Indonesia, the expectation is stranded, at least for two Dyaks of Kalimantan and two Sumatrans.

SUMMARY

Without trying to tabulate, it appears that most TEST specimens respond very well to assignment to a given main region, which is perhaps the best that should be expected. More specific assignments, e.g., Solomon Islands and New Hebrides as "Tolai," i.e., Melanesian, are even more satisfying but this is probably not really useful in the generality of cases. As expected, there is a proportion of blatantly "wrong" identifications, to remind us once more that these lie in wait in a small percentage of skulls being tested.

The above results probably give too favorable a picture, because the tested cases are in so many cases clearly related, geographically or ethnically, to some of the defining populations, e.g., the Polynesians or sub-Saharan specimens in table 4, and even including, for example, peoples in

North Asia who are not necessarily close to any such population. It is with the last lot above, designated Southeast Asia, that the problem is posed more acutely. I suspect that the same lurks in the Americas; there are hints that the good assignment in some cases (Peru) masks an unencompassed variation that is not properly tested.

Is the seeming lack of positive or expected assignment due to a) incomplete coverage of this Southeast Asian region by reference populations available, b) imperfections of method, or c) some lack of individuality in morphological character in the peoples concerned? This last may be an imaginary situation. In the old days such people might have been diagnosed as "mixed," i.e., not pure representatives of properly defined "races." That view really is of no help here.

4 Prehistoric Specimens

Here we come to a main object of the work. The skulls (or in many cases, casts) in table 5 are not a comprehensive list, but only those which became available during the course of measuring. They are also limited to specimens on which a complete set of the 57 measurements could be made, or to a certain degree estimated since, unlike the main body of material, it was often more important to include a specimen than to have it in a generally complete condition.

Late Prehistoric Europe

Nine of ten Portuguese Late Mesolithic skulls from Muge sites north of Lisbon are read, if not closely, as European, with one exception, thus a pretty positive statement. In fact, with the female preference for Zalavár, this could even be refined to "South European" or, in typological readings of old, "Mediterranean."

Kurgans are burial mounds widespread in Central Asia (see Anthony 1985). These specimens are not given a locality in the catalogue in Oslo, but were apparently gathered around the headwaters of the Yenesei River in eastern Siberia in the vicinity of Krasnoyarsk, and represent Scythians, seminomadic horse breeders of late prehistoric times. The affinities of both skulls are clearly European, a matter of interest in the ethnic relations of Scythians generally.

Four of five Danish Neolithic crania are also European in affiliation. The Ertebølle skull, however, is well removed from any modern population except Ainu. The Ainu, though Asiatic, are a population so generalized as not to be very indicative. In other analyses (Howells 1989), Ainu are apt to cluster with either Europeans or Guam.

This view is to be considered also in the case of the proto-Neolithic Natufian skull from the Levant. Ainu and Zalavár are the reasonable affiliations, which might be read as "generalized European."

Southern Africa and Australia

The Fingira 2 skull from the rock shelter of that name in Malawi is only about 3,000 years old. Its affiliations are clearly African, if remote; it is Bushman by primary indication.

The Fish Hoek skull from Peers Cave south of Cape Town has in the past been dated about 35,000 B.P. or more. Richard Klein gives me his opinion that the associations are doubtful, and that the likely date is end-Pleistocene, or about 10,000 B.P. Its assignment, here and by any other of these related assessments, is Bushman, quite

Who's Who in Skulls

TABLE 5-1

DISPOP results: Late Prehistoric Europe

1693 Muga	Zalvar 5.0 0.03	Norse 5.3 0.01	Egypt 5.7 0.00	S.Cruz 5.8 0.00	Peru 5.9 0.00	Ainu 5.9 0.00	Atayal 5.9 0.00	Berg 6.0 0.00	
1694 Muge	Hainan 4.7 0.07	Anyang 4.9 0.05	Atayal 4.9 0.04	Zalvar 5.0 0.03	Peru 5.0 0.03	S.Jap. 5.1 0.02	Norse 5.2 0.02	N.Jap. 5.4 0.01	
1695 Muge	Egypt 4.6 0.09	Andamn 5.1 0.02	Teita 5.2 0.01	Atayal 5.5 0.01	Zalvar 5.7 0.00	Hainan 5.8 0.00	S.Jap. 5.8 0.00	Norse 5.8 0.00	
1696 Muge	Zalvar 5.6 0.00	Atayal 5.7 0.00	Egypt 6.3 0.00	Teita 6.5 -	Hainan 6.6 -	Norse 6.6 -	Phil 6.6 -	Andamn 6.7 -	
1697 Muge	Egypt 4.3 0.17	Zalvar 4.5 0.12	Norse 4.6 0.09	Peru 5.4 0.01	Tolai 5.5 0.01	Berg 5.8 0.00	Tasman 5.9 0.00	Andamn 6.2 0.00	
1698 Muge		Zalvar 4.7 0.06	Norse 5.0 0.02	Egypt 5.1 0.02	Zulu 5.4 0.01	Easter 6.0 0.00	Berg 6.3 -	Tolai 6.4 -	Atayal 6.4 -
1699 Muge		Zalvar 5.1 0.02	Norse 5.2 0.01	Berg 5.5 0.00	Egypt 6.1 0.00	Arikar 6.4 -	Teita 6.8 -	Zulu 6.9 -	Tolai 6.9 -
1700 Muge		Zalvar 3.7 0.48	Egypt 3.8 0.40	Norse 3.8 0.39	Berg 4.9 0.03	Tolai 5.2 0.01	Austr 5.8 0.00	S.Cruz 5.9 0.00	Zulu 5.9 0.00
1701 Muge		Zalvar 3.5 0.57	Alevel 4.0 0.31	Ainu 4.6 0.07	Tasman 4.7 0.07	Egypt 4.7 0.07	Berg 4.7 0.06	Austr 4.7 0.06	Norse 4.0 0.05
1702 Muge		Zalvar 3.0 0.83	Norse 3.9 0.35	Tolai 4.2 0.20	Atayal 4.2 0.20	Berg 4.3 0.16	Austr 4.5 0.11	Egypt 4.5 0.10	Tasman 4.8 0.05
1559 Kurgan 3	Zalvar 2.7 0.90	Berg 3.7 0.46	Norse 4.1 0.27	Egypt 4.7 0.06	Peru 4.8 0.06	Ainu 4.8 0.05	Atayal 4.9 0.04	Phil 5.0 0.03	
1558 Kurgan 4	Norse 3.1 0.75	Zalvar 3.5 0.58	Egypt 3.9 0.35	Teita 4.8 0.05	Berg 4.9 0.04	S.Jap. 5.5 0.00	Hainan 5.8 0.00	Ainu 5.8 0.00	
1688 Borreby (Denmark)	Zalvar 3.7 0.45	Norse 4.5 0.12	Berg 4.7 0.07	Egypt 4.9 0.05	Hainan 5.3 0.01	S.Jap. 5.4 0.01	Atayal 5.5 0.01	Teita 5.5 0.00	
1689 Kyndeloese (Denmark)	Zalvar 3.9 0.35	Norse 4.1 0.25	Berg 4.1 0.24	Egypt 4.7 0.07	Ainu 4.9 0.04	S.Cruz 4.9 0.04	Tasman 5.1 0.02	Austr 5.7 0.00	
1690 Ertebolle (Denmark)	Ainu 5.1 0.02	N.Jap. 6.4 -	Egypt 6.6 -	S.Jap. 6.7 -	Dogon 6.8 -	Easter 7.0 -	Guam 7.0 -	Tolai 7.1 -	
1691 Neolithic (Denmark)	Egypt 3.7 0.44	Zalvar 4.1 0.26	Ainu 4.2 0.20	Andamn 4.5 0.12	Norse 4.5 0.11	Atayal 4.6 0.10	Zulu 4.6 0.09	S.Jap. 4.9 0.04	
1692 Neolithic (Denmark)	Zalvar 3.6 0.51	Egypt 3.9 0.34	Norse 4.2 0.20	Ainu 5.5 0.01	Berg 5.6 0.00	Zulu 6.4 -	Guam 6.4 -	Peru 6.5 -	
1775 Natufian	Ainu 4.6 0.10	Zalvar 4.6 0.08	Tasman 5.1 0.03	Atayal 5.1 0.02	Zulu 5.3 0.01	Austr 5.3 0.01	Phil 5.4 0.01	Berg 5.6 0.00	

TABLE 5-2
DISPOP results: Southern Africa and Australia

1926 Fingira 2		Bush	Teita	Zulu	Atayal	Andamn	S.Jap.	Tolai	Zalvar	
		5.7	7.2	7.9	7.9	8.1	8.3	8.5	8.5	
		0.00	–	–	–	–	–	–	–	
1703 Fish Hoek		Bush	Norse	Zalvar	S.Cruz	Berg	N.Jap.	Teita	Zulu	
		3.6	6.8	6.8	7.0	7.1	7.2	7.2	7.2	
		0.50	–	–	–	–	–	–	–	
1927 Matjes River	Cast	Bush	Egypt	Norse	Zalvar	Teita	Zulu	N.Jap.	Austr	
		4.4	6.4	6.5	6.8	6.8	7.0	7.3	7.3	
		0.14	–	–	–	–	–	–	–	
1704 Keilor		Tasman	Tolai	Easter	Arikar	Zalvar	Guam	Austr	Atayal	
		5.2	5.3	5.4	5.5	5.7	6.0	6.1	6.2	
		0.02	0.01	0.01	0.01	0.00	0.00	0.00	0.00	

positively. The same holds for a skull (actually a cast) from the Mytilus layer of the Matjes River rock shelter (near the Cape, to the east), with an age of 5,000 years or less. All three skulls appear solidly to confirm the accepted view of the Bushmen as the aboriginal population of southern Africa.

The Keilor skull from north of Melbourne, Australia, is 13,000 years old or more; this date is believed quite secure. Its assignment (not close) as Tolai or Tasmanian, rather than Australian, conforms to its general appearance, for whatever clues may be drawn as to local Australian population history. (The skull was measured before a recent cleaning of some surface encrustation.)

Prehistoric East Africa

Here are three sites in Kenya (Elmenteita, Nakuru, Willey's Kopje) about 50 miles northwest of Nairobi (see Leakey 1935). Leakey dated Nakuru and Willey's Kopje as "Neolithic" and the Elmenteita remains as "Early Mesolithic" (ca. 7000 B.C.). Coon (1962) accepted a Capsian culture and date and he diagnosed all the crania as having those same connections (i.e., North African), and as being Caucasoid in nature. In any case, he concluded, all antedate the Bantu expansion.

The Kenya "Capsian" culture is renamed Eburran (Rightmire 1984), and while similar dates prevail they are under some suspicion. Rightmire, with discriminant analysis, assigned Elmenteita A and B as most likely African Negro, not San, and emphatically not Caucasoid. In general, he sees no signs of population replacements in the area. The DISPOP results here are not indicative of anything, except a generally non-African nature for all these skulls. Display of POPKIN distances (infra) reinforces this and seems to find nearer neighbors among such more generalized populations as Peru, Guam, or Ainu, but also Europeans or even Easter Island.

Remembering that the Teita series (Bantu-speakers of southeastern Kenya), and the recent East African skulls in table 4 above, do clearly exhibit African affiliations, it is fair to say, contra Rightmire, that there seems to be no clear continuity here in late prehistory. On the broad scale, looking at an "Out-of-Africa" scenario, one would expect that, in some region between southern and northeastern Africa, some differentiation would have been taking place within a *Homo sapiens* stock, evolving into something beginning to approximate later sub-Saharan peoples on the one hand, and evolving in another direction on the other hand. East Africa would be a likely locale for appearance of the latter. So anyone is welcome to argue that this is what Elmenteita et al. are manifesting. The ensuing picture for East Africa, that is to say, would later have been changed through replacement by the expansion of Bantu or other "Negroid" tribes.

TABLE 5-3
DISPOP results: Prehistoric East Africa

1921 Elmenteita A	Austr	Arikar	Guam	Phil	Tolai	Teita	Peru	Hainan
	5.4	5.6	5.6	5.8	5.8	5.9	6.1	6.3
	0.01	0.00	0.00	0.00	0.00	0.00	0.00	0.00
1922 Elmenteita B	Guam	Hawaii	Peru	Egypt	Norse	S.Jap.	Phil	S.Cruz
	4.6	5.0	5.1	5.1	5.2	5.3	5.4	5.5
	0.09	0.03	0.02	0.02	0.02	0.01	0.01	0.01
1923 Nakuru 9	Austr	Egypt	Peru	Norse	Teita	Zalvar	Zulu	Tolai
	6.0	6.1	6.1	6.2	6.4	6.7	6.7	6.7
	0.00	0.00	0.00	0.00	-	-	-	-
1925 Willey's Kopje 3	Guam	Easter	Ainu	Zalvar	Zulu	S.Jap.	Atayal	Norse
	5.0	5.0	5.0	5.2	5.3	5.4	5.5	5.8
	0.03	0.03	0.03	0.02	0.01	0.01	0.00	0.00

TABLE 5-4
DISPOP results: End-Pleistocene/Holocene

2076 Afalou 5	Phil	Tasman	Ainu	Guam	N.Jap.	Austr	Tolai	Zalvar
	4.5	4.9	4.9	5.0	5.1	5.4	5.4	5.5
	0.12	0.04	0.04	0.03	0.02	0.01	0.01	0.01
2077 Afalou 9	Tasman	Zalvar	Berg	Tolai	Austr	Phil	S.Cruz	Norse
	6.0	6.0	6.2	6.0	6.3	6.5	6.5	6.5
	0.04	0.03	0.02	0.00	0.00	-	-	-
2078 Teviec 11	Easter	Hawaii	Guam	Tolai	Ainu	Zalvar	Tasman	Morior
	3.8	4.7	5.8	5.8	6.0	6.2	6.2	6.4
	0.41	0.07	0.00	0.00	0.00	0.00	0.00	-
2079 Teviec 16	Zalvar	Ainu	N.Jap.	S.Jap.	Arikar	Atayal	Hainan	Phil
	4.5	4.6	4.8	5.1	5.1	5.2	5.2	5.3
	0.11	0.08	0.05	0.03	0.02	0.02	0.02	0.01

End-Pleistocene/Holocene

The remains from Afalou-bou-Rhummel, Algeria, are accompanied by an early Oranian culture, with microlithic blades. Lahr (1992) quotes dates of 14,000 to 8500 B.C. Coon (1939) considered this series in detail and, like others before him, found a close correspondence with Upper Paleolithic Europeans (the North Africans being even more robust), resembling in particular "western Cro Magnons." The nose, Coon stated, is European in bony configuration, although nose and face are relatively not so long. However, affiliations by DISPOP are obscure; a clear European diagnosis is not forthcoming. Afalou 5 emerges as "Philippines," an odd place for something so robust, except that Philippines, as noted

above, seems like the most protean among the series and thus a refuge for crania not closely affiliated otherwise. In DISPOP using principal components (not shown herein) rather than canonical variates, Afalou 5 appears as Ainu, so a similar interpretation may be suggested. In POPKIN (infra) the skull also has Philippines and Guam as nearest neighbors, with a few Zulu; further nearest neighbors are North Japan, Zalavár, and Ainu.

Afalou 9 has a slightly better claim to be European, but also Tasmanian (the Afalou orbits are particularly low). The problem is low probability for all. By nearest neighbors (POPKIN, below), Zalavár wins hands down, followed by Berg and Tasmania.

Two specimens come from the Mesolithic site at Teviec, an island off Brittany. Although the culture is Mesolithic, the date must be relatively late (fourth millennium B.C.). By Coon's report (1939) the skulls are moderately robust, though diminished relative to Upper Paleolithic European skulls (even relative to Muge, he avers, which is contrary to my own impression of gracility in the Muge lot). Number 16 is "European." Number 11 cannot reasonably have most of the suggested assignments; however, the insistent Polynesian reference for this skull may suggest its morphological nature as something more rugged than later Europeans. Of course, this exercise is meant to find relations to modern peoples, not morphological descriptions.

This brings up an important consideration not directly uncovered in the present study. The same apparent displacement of late prehistoric specimens relative to moderns may be seen in a recent study by Marta Lahr (1992). This is an intensive comparison, using metric and non-metric characters, of skull samples from five world regions, adding also the complete set of Afalou and contemporary Taforalt (Morocco) specimens, numbering 59 in all. She finds her composite Australo-Melanesian sample to be somewhat set apart from other regions by greater robustness, while the prehistoric North Africans are still more differentiated, by larger size, greater height and robustness, and larger teeth. They stand outside modern variation, though she raises the key question of whether modern variation should be broadened to include them, accepting their robustness.

In the present analysis Australoid size or robusticity is supposedly vitiated by the use of C-scores, and this would apply to the Afalou, Teviec, and Elmenteita materials as well. Accordingly, from Lahr's work, these Mesolithic or Epi-Paleolithic specimens apparently should be viewed as having a character, roughly taken as "robustness," that is not encompassed in the variation defined by our modern samples and makes obscure their relations to recent populations, real though such connections may be. This situation must be borne in mind in considering all Late Pleistocene specimens herein.

Upper Cave 101

Upper Cave 101 from Zhoukoudian is Late Pleistocene, currently dated at 25,000 B.P. or earlier (Hedges et al. 1992), at a time of relatively mild climate. This specimen is everyone's problem skull, still the subject of discussion as to its ethnic affiliation. Earlier, when its estimated date was much more recent, the question was whether any connection could be seen between Upper Cave 101 and the early Chinese of about 7000 B.C. (Howells 1983), because of the evident disparity in morphology. Usually neglected, because of its damaged condition, was Upper Cave 103, clearly more "Mongoloid" in general appearance.

In Weidenreich's opinion (1939), Upper Cave 101 was "primitive Mongolian." Chinese and some other scholars have usually seen it as situated on a developmental line from earlier populations to modern Chinese, but such local affiliation has not been broadly agreed upon. G. Neumann (1959), experienced in cranial morphological matters, believed the skull's closest relations were with American Indians. I could not find univariate support for this (1989), but instead found its nearest C-score likenesses to be with Polynesians.

The skull was the principal object of R. Wright's formulation of CRANID (see below), on the basis of which he and Kamminga (1988) plotted Upper Cave 101 (on second and third principal components) in a multivariate area well separated from a block of Mongoloids/Americans/Polynesians on one hand and lying between one of Europeans and one of Africans on the other, with Ainus as the nearest population. (They took most of the same measurements as in the present work; they measured a different cast—no originals survive—but arrived at very similar measurements.) In his latest examinations (1992b), Wright locates the skull closest to Tolais and Australians, followed by Tasmanians and Bushmen. His analyses are multiplex, but I think this would argue the total exclusion of any close "Mongoloid" connections by these methods for the skull.

Van Vark and Dijkema (1988) carried out a full D^2 testing of distances of Upper Cave 101 from all of the populations used herein, including

TABLE 5-5
DISPOP results: Upper Cave 101

2718 Upper Cave 101 Cast	Arikar	Zalvar	Easter	Norse	Berg	Tasman	Morior	Hawaii
	6.0	6.1	6.2	6.2	6.2	6.3	6.8	7.0
	0.00	0.00	0.00	0.00	0.00	0.00	–	–

both sexes. In every case the skull was out of the modern range, though lying closest to Australians but also to Norse, Ainu, Arikara, Santa Cruz, and Easter Island. In no case were Asiatic series close. Except for Australians, this distribution is close to mine, below.

David Bulbeck (1981) has done a similar study focusing on Asiatic materials, finding similar placements, especially in seeing no Asiatic home for UC 101. In his work UC 101 comes to rest in a space of its own. He suggests (by letter) that UC 101 is to later Mongoloids as Wajak is to Indonesians. This is an apt comment. However, Wajak, which is not analyzed herein, seems visually to be an Australoid outlier rather than to approach recent Indonesians, which is how the Multiregional Evolution School places it in their course from *Homo erectus* to modern Australians in this region.

Brace (n.d.), with different measurement sets and reference samples, used discriminant functions to rate probabilities of group membership, finding that Upper Cave 101 would not be excluded from his Jomon-Pacific grouping. Because he uses posterior probabilities, this does not really allow such membership, as explained earlier. In another such test (Brace and Tracer 1992) with the same questionable usage, he finds that UC 101 might be a member of either his Pacific or North/East Amerind groups.

DISPOP, herewith, agrees in denying any North Asian connections for Upper Cave 101. It does, however, consider Arikara as least distant, though all populations are distant to a highly significant degree. (As related earlier I have carried out another version of DISPOP, using principal components rather that canonical variates, which should parallel Wright's CRANID, which also uses principal components. In any event, the results approximate those of table 5-5 rather than those of CRANID.)

To anticipate, inspection of POPKIN (infra) also places Arikara as closest to UC 101 on all counts, with all 30 Arikara falling in the nearest moiety. Norse and Zalavár have a good number of low distances, more than Berg. Easter Island is probably next. Most distant specimens are Africans and North Asians. Indians other than Arikara are not at all close. Tolai are intermediate. Australians and Tasmanians are not close.

The skull bears all this inspection because of its importance and because so much attention has been paid to it. Following methods presented here, Upper Cave 101 fits most nearly in a cluster which includes Europeans and American Indians, a branch which appears in clusterings shown herein (fig. 3). I think that this is the most judicious assessment of the morphology of the skull, and one which is not in gross disagreement with what others have found. At the same time we should keep in mind the stricture introduced above, as suggested by the work of Lahr with the North African material. Skulls of the late Pleistocene or early Holocene transition, recent though they are, may have a general character of robustness that makes them resist allocation to fully modern populations.

This position agrees with one arrived at by Lahr, in another new study (n.d.) dealing with cranial diversification in American Indian populations. From their lack of close identification with Asiatic populations (fig. 3 herein) she proposes that some trans-Bering movements, especially earlier ones, were drawn from "pre-Mongoloid" populations of eastern Asia, i.e., before the emergence of the "typical" Mongoloids of recent times in North China and northward. This would envisage a protean North Eurasiatic population in which the Upper Cave sample would be at home, as the figures suggest.

TABLE 5-6
DISPOP results: Upper Paleolithic Europe

1705 Chancelade Cast	Egypt 8.5 –	Norse 8.7 –	Teita 8.9 –	Easter 9.3 –	Zalvar 9.3 –	Eskimo 9.9 –	Ainu 10.0 –	Tolai 10.1 –	
1706 Predmosti 3 Cast	Norse 7.2 –	Zalvar 7.3 –	Egypt 7.6 –	Tasman 7.7 –	Morior 7.9 –	Berg 7.9 –	Hawaii 8.3 –	Austr 8.3 –	
1707 Predmosti 4 Cast	Tolai 7.0 –	Zalvar 7.2 –	Tasman 7.3 –	Norse 7.6 –	Austr 8.0 –	Berg 8.2 –	Zulu 8.5 –	Egypt 8.5 –	
1707 Predmosti 4 Cast	Egypt 5.4 0.01	Zalvar 5.4 0.01	Norse 5.6 0.00	Tasman 6.0 0.00	Austr 6.4 –	Tolai 6.5 –	Berg 6.6 –	Ainu 6.7 –	
1708 Markina Gora (Kostenki)	Tasman 4.2 0.22	Austr 5.3 0.01	Tolai 5.5 0.01	Zalvar 5.7 0.00	Berg 6.4 –	Norse 6.4 –	Egypt 6.6 –	Ainu 6.9 –	
2716 Mladec 1 Cast	Zalvar 4.1 0.27	Norse 4.6 0.08	Tasman 4.7 0.07	Berg 4.8 0.06	Tolai 5.2 0.01	Egypt 5.5 0.00	Austr 5.6 0.00	Teita 5.9 0.00	
2904 Abri Pataud	Norse 4.3 0.19	Zalvar 5.3 0.01	Teita 5.4 0.01	Arikar 5.4 0.01	Zulu 5.5 0.00	Berg 5.6 0.00	Tolai 5.6 0.00	Egypt 5.7 0.00	

Upper Paleolithic Europe

Here the question is the filiation of recent Europeans from those of the Upper Paleolithic. There has existed a general appreciation of such a connection, on the basis of cranial character and appearance, but with varied qualifications. Some, like Brace (e.g., 1991), have noted the greater robustness and larger teeth of the Upper Paleolithic people. Others have looked for this, and for other perceived characteristics, in some kind of derivation from Neanderthals, or from the Skhūl people, which to Coon (1939) was not a great distinction. For others still, e.g., Fred Smith (1984), the Neanderthal connection has been one of gene flow from Neanderthals. Recently van Vark, Bilsborough, and Henke (1992; see further on) have used special tests of significance of differences among D^2 distances on these questions. Using an omnium gatherum of Late Paleolithic European skulls, they find this Upper Paleolithic set to be significantly more distant from modern Europeans than from an intermediate Mesolithic sample. This backward tracking, however, does not make the earlier group more similar to Neanderthals. The latter, they find, are more distant

from Europeans, early or late, than from some other modern populations.

The only questions that can be asked here are whether Upper Paleolithic specimens are closer to Europeans than to other recent samples, and also whether anything can be judged regarding Neanderthal distances from both.

Předmostí 3 was originally diagnosed as male, Předmostí 4 as female. (Only casts are available, the originals having been destroyed in World War II.) If Předmostí 4 is tested as male, it is within striking distance of Egypt and Zalavár; but Předmostí 3 and Chancelade, although also finding such populations as nearest neighbors, are very distant from all populations, so that any approaches to Europeans are probably meaningless. Of course, Neanderthal characters have been perceived in the Předmostí crania, but not in Chancelade, which by earlier writers was commonly diagnosed as Eskimoid.

Mladeč 1 (early Upper Paleolithic of the Czech Republic) and Abri Pataud (Les Eyzies in the Dordogne) are both comfortably European. Markina Gora (Kostenki in southern Russia) has strong but not heavy brows, and this brow structure may somehow lead to the possible assignment as Tasmanian.

TABLE 5-7

DISPOP results: Earlier Pleistocene

2083 Djebel Quafzeh 6	Tasman	Austr	Tolai	Ainu	Zulu	Guam	Morior	Hawaii	
	4.7	5.5	5.5	5.7	6.2	6.2	6.2	6.4	
Orig.	0.08	0.01	0.01	0.00	0.00	0.00	0.00	–	

1772 Skhul 5	Tasman	Austr	Tolai	S.Cruz	Phil	Arikar	N.Jap.	Ainu	
Orig.	5.6	7.3	7.4	8.8	8.9	9.0	9.0	9.1	
	0.00	–	–	–	–	–	–	–	

1773 Skhul 5	Tasman	Tolai	Austr	S.Cruz	N.Jap.	Arikar	Phil	Zulu	
Cast	7.1	8.3	8.9	9.8	10.0	10.1	10.1	10.5	
	–	–	–	–	–	–	–	–	

2080 Djebel Irhoud 1	Tasman	Austr	Tolai	S.Cruz	Arikar	Zulu	Berg	Bush	
Cast	8.1	9.2	9.8	10.3	10.3	10.5	10.6	10.6	
	–	–	–	–	–	–	–	–	

1778 Shanidar	Arikar	Berg	Tasman	Morior	Zalvar	Norse	S.Cruz	N.Jap.	
Cast	7.7	8.1	8.2	8.3	8.6	8.7	9.0	9.3	
	–	–	–	–	–	–	–	–	

2082 La Chapelle	Tasman	Austr	Morior	S.Cruz	Tolai	Berg	N.Jap.	Arikar	
Orig.	9.1	10.3	10.8	10.9	10.9	10.9	11.0	11.0	
	–	–	–	–	–	–	–	–	

2903 La Ferrassie	Tasman	Arikar	Morior	S.Cruz	Berg	Austr	Norse	Zalvar	
Cast	7.0	7.2	7.3	7.4	7.6	7.8	8.0	8.1	
	–	–	–	–	–	–	–	–	

1713 Steinheim	Tasman	Tolai	Austr	Easter	Berg	Arikar	Morior	Peru
	10.9	11.3	12.5	13.7	14.2	14.3	14.4	14.4
	–	–	–	–	–	–	–	–

1712 Broken Hill T	Austr	Tasman	S.Cruz	Tolai	Arikar	Morior	Zulu	Bush	
Orig.	8.3	8.3	9.1	9.6	9.7	9.8	9.9	9.9	
	–	–	–	–	–	–	–	–	

Earlier Pleistocene

The same may be responsible for the very similar distances for Qafzeh 6, whose brow structure is the least "modern" aspect of that cranium and also of the two versions of Skhūl 5. The divergence of the very good cast of Skhūl 5 from the original skull is a caution.

Jebel Irhoud 1 (also a cast), usually described as near-modern (and by most writers as non-Neanderthal) is over 7 standard deviations from any population. The same is true of all Neanderthal representatives and of Broken Hill, with a cast of Steinheim being extremely distant from moderns. This indicates the non-feasibility of trying to connect such fossils with any living people.

5 POPKIN

This second method attempts to place a target skull in a universe of known individuals. These individuals are not grouped by population, but rather arranged in toto by their mutual distances, without regard to population membership (Howells 1986). The ethnic affiliation of the target skull is suggested by the neighborhood in which it finds itself, read from its nearest neighbors, now identified by population. As mentioned earlier, this is analogous to Wright's CRANID.

PRUNED SAMPLES

In order to simplify the whole procedure, with regard to computation and in order to equalize the chances among populations for providing near neighbors, the 28 *male* series were reduced to 30 members each (except for Atayals, who number only 29). This gives a total matrix of 839 skulls. (In DISPOP, where only group means were used, the full available series were kept so as to use all possible information.) The female series, mostly smaller in numbers, are not considered here.

In addition to the reasons above, there are other advantages in using samples of equal numbers. (See Albrecht 1992.)

A total of 30 cases per series is doubtless ample to represent a population, for all present purposes. Knußman (1967), investigating some problems of sample sizes in multivariate analysis, concluded that "large" samples of the order of N=25

were desirable. Here, as an experiment, population clusterings were run of randomly selected samples of sizes 1, 2, 3, 5, 7, 10, and 15 from each population, over 10 separate selections for each sample size. The clusterings in fact became relatively stable over the 10 runs at a sample size of only 3. The clusterings varied in detail, as they do with larger samples, but the configurations were much the same, and conformed to expectation. Naturally, differences among a set of 10 clusterings became smaller as sample size increased, and were almost invariant at a sample size of 15.

As used, the final pruning was performed by a single random selection of 30 from each population (always excepting the 29 Atayals).

Logically, such pruning could have been done in various ways. One obvious method was attempted. Using C-scores, the distance from its appropriate group centroid as computed for each individual, in each group those most distant from the centroid were culled until 30 were left. The aim here was to render the surviving groups mutually more discrete in a multivariate space. Such was the actual result, since culled individuals, and others assignable to the same population (like the extra Peruvians in the TEST series, described earlier under DISPOP) tended to have an overall distribution of distances varying further from the mean than that of members of the pruned group. However, an inherent unfavorable result is the diminishing of within-group variation, and in this way artificially increasing distances from a group centroid of actually "normal" members of the group, and thus lowering their computed probability of membership. On experiment, that turned out to be the case here; accordingly, this method of pruning was rejected.

The method used, random pruning, appears to have obviated these objections. There resulted smaller overall differences between "prunes" and "culls" in their distances from group means, and more expectable probabilities of group membership for the "culls," those group members who had been removed purely at random. Figure 2 illustrates the effect of the two kinds of pruning on the Peruvian population, and the advantages of random pruning.

As a final note, legitimacy of random pruning was tested by performing a second such pruning, which gave results close to, but not identical with the first. This suggests that a) any such pruned set is as good as any other; b) any such is no worse than pretending that only 30 skulls were available in collections; and c) readings from POPKIN look arithmetically more exact than they really are.

Centrality of Shape

For any individual, the mean of C-scores is zero, by definition. The C-scores are distributed around zero, plus and minus, and have a standard deviation. The formal definition of a standard deviation is, of course, the square root of the mean of the squared deviations. This is the same as the definition, here, for the Euclidean distance of an individual's zero mean from a grand, unseen zero mean of all individuals seen as

shape. That is to say, in producing C-scores, the entire set of raw measurements was originally standardized, for an individual, on a zero mean and a unit standard deviation, and then, in each case, re-standardized to C-scores by the subtraction of PENSIZE. The remaining deviations from this new zero thus become strictly those of shape. And the square root of their mean squares, while serving as a standard deviation of an individual's C-scores, also constitutes a *shape* distance from the central tendency shape for the whole universe.

Table 6 gives the mean of the C-score standard deviations for the 30 members of each series (as established by random pruning.) This is meant to show which populations are most or least extreme in shape departures from a central shape for this whole universe. It is *not* a measure of variability for a population. It may be seen that the figure for Tasmanians does not reflect the especially high variation within the Tasmanian series seen in table 1. And it does *not* imply that some populations represent a form from which others have departed, an *Urtyp* which is in any way parental to others.

TABLE 6
Mean C-score standard deviations

Atayal	29	9.68
Zalavar	30	9.78
Philippines	30	9.79
Guam	30	9.82
Santa Cruz	30	9.84
N. Japan	30	9.98
Ainu	30	9.98
S. Japan	30	10.02
Arikara	30	10.12
Hainan	30	10.14
Peru	30	10.30
Norse	30	10.32
Moriori	30	10.49
Tolai	30	10.58
Hawaii	30	10.76
Anyang	30	10.76
Zulu	30	10.89
Tasmania	30	10.96
Egypt	30	11.02
Andaman	30	11.13
Dogon	30	11.40
Australia	30	11.49
Easter I	30	11.49
Berg	30	11.60
Teita	30	11.64
Bush	30	12.16
Eskimo	30	13.36
Buriat	30	15.19

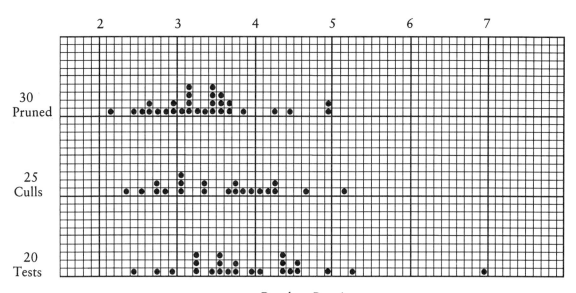

Random Pruning

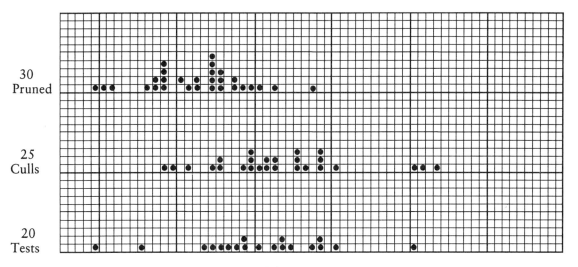

Selective Pruning

Figure 2. The "pruned" group comprises 30 skulls, out of the 55 of the male base series, after culling 25 skulls, either by random selection (above) or by removing those furthest from the multivariate group mean. The "tests" are 20 skulls (of the 75 originally measured) which were removed randomly from the base series before any work.

Shown are distributions of individual distances from the group mean. It is apparent a) that the selectively pruned distribution is slightly closer to the mean than that of the randomly pruned set, and b) that with random pruning, both culls and tests are only slightly displaced toward greater distance compared to selective pruning. The net effect would be to reject as "normal" cases more of the latter distributions.

The information is simply that some populations are more extreme in shape, such as Buriats above all, also Eskimo, Bush, Teita, Berg—whether these are long- or short-headed, long- or short-faced—and that others (often Asiatics), such as Atayals, Zalavár, Filipinos, are less distinguished in aspects of form. This is not intrinsically interesting information; it could have been made out from the detailed C-scores in *Skull Shapes*. It is shown here as a statistic for one reason. When, in tests for proximity of a target skull, there is less often a nearness to Buriats, let us say, and more often one to Atayals, Zalavárs, or Filipinos, this likelihood may constitute an extra element of uncertainty, and does not constitute an indication that there has been an Atayal Eve (or Adam—we are all males here). For purposes of interpretation of POPKIN, all this may bear on the fact, seen below, that Eskimo and Buriat individuals are usually well removed in position from a target skull.

Population Relationships

Table 7 gives the Euclidean distances among the mean positions of the 28 pruned groups, based on 14 canonical variates (see above). This shows expectable results, especially with regard to there being smaller distances within major geographical groups. Exceptions are rare: Teita is very slightly further from Dogon than from Egypt or Bush, not an impressive discrepancy. Easter Island, likewise, is very slightly closer to

Anyang, Guam or Atayals than to Moriori, here again a straying that does not impress.

Figure 3 shows the same results in dendrogram form. This is very close indeed to Figure 15 in *Skull Shapes* (reproduced herein as fig. 7) which was formed from Mahalanobis' D based on all possible canonical variates, and on total, not pruned, samples.

Table 8 is a somewhat different set of distances. This is computed as Steerneman's "Type 1" D^2 distance. In the diagonal cells there appears the mean, for the group concerned, of all the distances of each *individual* from every other member of the group concerned, a total of 435 distances for each group. The off-diagonal cells contain the mean of the distances of each member of one group from each member of the other group, the total for each cell being 900 (from 30 x 30). In all cases, of course, there is a reduction in the total where the 29 Atayals are involved. (These hundreds of thousands of distances were computed with a program developed and kindly provided by A. G. Steerneman of Groningen University.) This is the most specific possible expression of within-group and between-group differences.

This also allows the computation of the mean D^2 of all the intragroup distances, i.e., all the inter-individual distances included in the diagonal cells of table 8. The total number of such distances is 12,151. The mean D^2 is 28.0167 and the mean D is thus 5.29. This can be used as a gauge of the POPKIN values in the tables following.

POPKIN TABULATIONS

These are too voluminous to allow publication of all the target cases appearing in tables 4 and 5, under DISPOP. As a first example of a POPKIN display we may take #1426, a Peruvian TEST case. This appears as good average "Peruvian;" in table 2 it had a "Peru" distance of 3.6 (thus at the mean of 3.62 for all DISPOP distances) and a "Peru" probability of .50. (Next nearest population means were Zalavár, D=4.9, P=.04; and Santa Cruz, D=5.3, P=.01.)

Table 9A-1 shows all 839 of the control (pruned) set, with the rank of each in its distance from #1426. Each is slotted into its appropriate population, according to rank in the total, with its Euclidean distance in the second line. The 30 in each line are parted into equal halves by a separation, which thus marks the median rank of distances for that population.

The right-hand column gives the mean rank for that line, above, and the mean distance, below. Note that this mean distance is quite close to the

TABLE 7
Distances among pruned population samples

	Norse	Zalav	Berg	Egypt	Teita	Dogon	Zulu	Bush	SAust	Tasman	Tolai	Mokapu	Easter	Morior
Norse	-	2.4	3.5	3.1	5.6	7.3	6.1	7.0	6.6	6.8	6.3	7.4	7.4	6.2
Zalavar	2.4	-	2.7	3.4	5.3	6.9	5.0	6.4	5.9	5.9	5.6	6.9	7.1	6.9
Berg	3.5	2.7	-	5.0	6.5	7.7	6.1	6.5	7.1	6.4	6.8	7.7	7.9	7.5
Egypt	3.1	3.4	5.0	-	5.1	5.9	5.2	7.3	6.8	7.1	6.3	7.5	7.8	6.7
Teita	5.6	5.3	6.5	5.1	-	5.3	3.9	5.2	6.2	7.2	5.7	7.9	6.9	7.6
Dogon	7.3	6.9	7.7	5.9	5.3	-	3.8	6.8	7.6	7.9	6.7	8.3	8.4	8.5
Zulu	6.1	5.0	6.1	5.2	3.9	3.8	-	5.7	6.0	6.5	5.6	7.2	6.9	7.6
Bushman	7.0	6.4	6.5	7.3	5.2	6.8	5.7	-	7.4	7.5	7.5	9.7	9.0	9.7
Australia	6.6	5.9	7.1	6.8	6.2	7.6	6.0	7.4	-	4.7	4.8	8.9	8.9	8.6
Tasmania	6.8	5.9	6.4	7.1	7.2	7.9	6.5	7.5	4.7	-	3.7	7.3	7.3	7.6
Tolai	6.3	5.6	6.8	6.3	5.7	6.7	5.6	7.5	4.8	3.7	-	6.3	6.0	6.8
Hawaii	7.4	6.9	7.7	7.5	7.9	8.3	7.2	9.7	8.9	7.3	6.3	-	4.0	4.8
Easter I	7.4	7.1	7.9	7.8	6.9	8.4	6.9	9.0	8.9	7.3	6.0	4.0	-	5.7
Moriori	6.2	6.9	7.5	6.7	7.6	8.5	7.6	9.7	8.6	7.6	6.8	4.8	5.7	-
Arikara	5.4	5.2	5.1	6.5	6.9	8.6	7.2	8.8	7.6	6.8	6.3	6.1	6.7	5.3
Santa Cruz	5.0	4.7	4.9	5.7	6.8	8.0	6.7	7.5	6.1	6.2	5.5	7.2	8.1	6.1
Peru	4.7	4.4	5.0	5.6	6.6	7.4	6.4	8.1	7.1	6.8	5.2	6.1	6.7	5.9
N Japan	6.0	5.3	6.0	6.0	6.3	6.4	5.1	7.4	8.0	7.3	5.7	5.7	6.0	5.6
S Japan	6.5	5.8	6.6	6.1	5.9	6.5	5.3	7.6	8.8	8.2	6.5	5.9	5.9	6.2
Hainan	5.9	5.3	5.9	5.9	6.1	6.1	5.6	7.5	8.4	7.5	6.1	5.8	5.8	6.2
Anyang	6.6	6.2	7.0	6.9	6.4	7.2	6.4	7.7	9.0	8.0	6.6	5.5	5.1	5.8
Atayal	6.0	5.0	5.4	6.2	5.6	6.0	4.9	6.5	7.7	7.1	5.8	6.5	5.6	7.1
Philippines	6.0	4.8	5.7	5.6	5.7	5.6	4.6	6.7	7.5	6.5	5.8	5.5	6.1	6.4
Guam	6.2	5.4	6.8	6.1	5.9	6.6	5.8	8.1	7.2	7.2	5.6	4.4	5.4	6.1
Ainu	4.8	4.6	5.6	5.0	6.2	6.1	5.2	7.4	6.6	6.9	6.1	6.3	6.1	5.7
Andaman Is	6.2	6.0	6.3	5.1	5.8	5.4	5.7	6.5	7.8	6.9	6.3	7.2	7.6	7.2
Buriat	7.6	7.2	6.3	8.8	8.7	9.6	8.7	8.9	10.4	9.7	9.2	8.5	9.5	8.7
Eskimo	6.9	6.6	7.5	7.4	6.6	9.3	7.9	8.9	7.9	8.8	6.6	7.7	7.4	8.0

	Arikar	S Cruz	Peru	NJapan	SJapan	Hainan	Anyang	Atayal	Philip	Guam	Ainu	Andamn	Buriat	Eskimo
Norse	5.4	5.0	4.7	6.0	6.5	5.9	6.6	6.0	6.0	6.2	4.8	6.2	7.6	6.9
Zalavar	5.2	4.7	4.4	5.3	5.8	5.3	6.2	5.0	4.8	5.4	4.6	6.0	7.2	6.6
Berg	5.1	4.9	5.0	6.0	6.6	5.9	7.0	5.4	5.7	6.8	5.6	6.3	6.3	7.5
Egypt	6.5	5.7	5.6	6.0	6.1	5.9	6.9	6.2	5.6	6.1	5.0	5.1	8.8	7.4
Teita	6.9	6.8	6.6	6.3	5.9	6.1	6.4	5.6	5.7	5.9	6.2	5.8	8.7	6.6
Dogon	8.6	8.0	7.4	6.4	6.1	6.1	7.2	6.0	5.6	6.6	6.1	5.4	9.6	9.3
Zulu	7.2	6.7	6.4	5.1	5.3	5.6	6.4	4.9	4.6	5.8	5.2	5.7	8.7	7.9
Bushman	8.8	7.5	8.1	7.4	7.6	7.5	7.7	6.5	6.7	8.1	7.4	6.5	8.9	8.9
Australia	7.6	6.1	7.1	8.0	8.8	8.4	9.0	7.7	7.5	7.2	6.6	7.8	10.4	7.9
Tasmania	6.8	6.2	6.8	7.3	8.2	7.5	8.0	7.1	6.5	7.2	6.9	6.9	9.7	8.8
Tolai	6.3	5.5	5.2	5.7	6.5	6.1	6.6	5.8	5.8	5.6	6.1	6.3	9.2	6.6
Hawaii	6.1	7.2	6.1	5.7	5.9	5.8	5.5	6.5	5.5	4.4	6.3	7.2	8.5	7.7
Easter I	6.7	8.1	6.7	6.0	5.9	5.8	5.1	5.6	6.1	5.4	6.1	7.6	9.5	7.4
Moriori	5.3	6.1	5.9	5.6	6.2	6.2	5.8	7.1	6.4	6.1	5.7	7.2	8.7	8.0
Arikara	-	4.1	3.9	5.7	5.7	4.8	5.8	5.3	5.2	5.4	6.3	6.1	6.7	7.0
Santa Cruz	4.1	-	3.5	5.1	5.9	5.6	6.6	5.6	5.5	6.1	5.9	5.4	7.8	6.8
Peru	3.9	3.5	-	4.7	5.2	4.2	5.4	4.5	4.9	5.1	5.9	5.6	7.8	7.2
N Japan	5.7	5.1	4.7	-	2.2	3.1	3.5	3.6	3.4	4.6	4.3	5.9	7.0	6.5
S Japan	5.7	5.9	5.2	2.2	-	2.4	2.9	3.3	3.2	4.3	5.1	5.5	7.2	6.4
Hainan	4.8	5.6	4.2	3.1	2.4	-	2.4	2.3	2.4	3.7	4.8	4.9	6.9	7.0
Anyang	5.8	6.6	5.4	3.5	2.9	2.4	-	3.7	3.4	3.9	5.1	6.3	7.2	6.9
Atayal	5.3	5.6	4.5	3.6	3.3	2.3	3.7	-	3.2	4.6	4.7	5.0	7.6	7.0
Philippines	5.2	5.5	4.9	3.4	3.2	2.4	3.4	3.2	-	3.5	4.8	4.8	7.0	7.5
Guam	5.4	6.1	5.1	4.6	4.3	3.7	3.9	4.6	3.5	-	4.7	6.1	7.2	5.7
Ainu	6.3	5.9	5.9	4.3	5.1	4.8	5.1	4.7	4.8	4.7	-	6.2	7.9	6.5
Andaman Is	6.1	5.4	5.6	5.9	5.5	4.9	6.3	5.0	4.8	6.1	6.2	-	9.2	8.4
Buriat	6.7	7.8	7.8	7.0	7.2	6.9	7.2	7.6	7.0	7.2	7.9	9.2	-	7.5
Eskimo	7.0	6.8	7.2	6.5	6.4	7.0	6.9	7.0	7.5	5.7	6.5	8.4	7.5	-

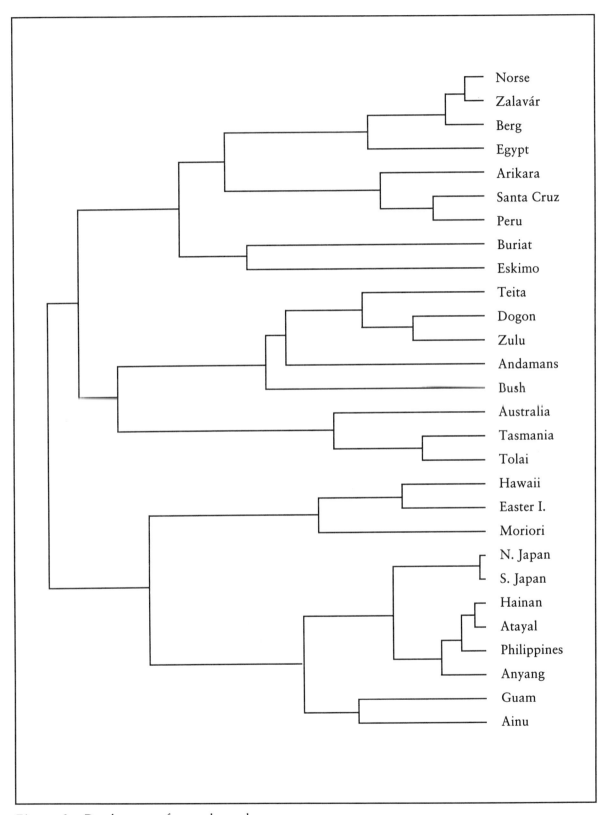

Figure 3. Dendrogram of pruned samples.

TABLE 8

Means of cross-distances between all individuals of groups shown

	Norse	Zalvar	Berg	Egypt	Teita	Dogon	Zulu	Bush	S.Aust	Tasman	Tolai	Mokapu	Easter	Morior
Norse	27.4	30.6	39.1	35.0	60.5	81.7	65.6	77.2	69.6	78.1	65.5	81.3	81.0	65.0
Zalvar	30.6	24.3	33.0	35.4	56.3	74.9	51.9	66.5	59.0	65.4	56.7	73.3	75.0	72.9
Berg	39.1	33.0	28.9	51.6	73.2	88.6	65.4	70.1	77.1	73.9	72.9	86.0	89.6	83.6
Egypt	35.0	35.4	51.6	25.5	54.2	62.5	53.6	79.3	70.6	81.0	65.3	81.4	85.7	71.0
Teita	60.5	56.3	73.2	54.2	33.8	60.4	46.1	58.0	67.5	85.9	61.9	91.8	77.0	87.8
Dogon	81.7	74.9	88.6	62.5	60.4	32.7	44.8	76.2	85.3	96.9	73.3	98.3	99.1	102.3
Zulu	65.6	51.9	65.4	53.6	46.1	44.8	30.3	60.9	63.3	75.2	58.7	80.3	76.0	86.5
Bush	77.2	66.5	70.1	79.3	58.0	76.2	60.9	29.6	81.3	88.4	84.1	121.2	109.4	122.6
S.Aust	69.6	59.0	77.1	70.6	67.5	85.3	63.3	81.3	25.7	53.1	48.3	104.7	104.3	99.7
Tasman	78.1	65.4	73.9	81.0	85.9	96.9	75.2	88.4	53.1	38.1	45.5	85.0	85.4	90.2
Tolai	65.5	56.7	72.9	65.3	61.9	73.3	58.7	84.1	48.3	45.5	27.3	65.9	62.8	72.9
Mokapu	81.3	73.3	86.0	81.4	91.8	98.3	80.3	121.2	104.7	85.0	65.9	27.9	42.7	50.4
Easter	81.0	75.0	89.6	85.7	77.0	99.1	76.0	109.4	104.3	85.4	62.8	42.7	27.6	59.6
Morior	65.0	72.9	83.6	71.0	87.8	102.3	86.5	122.6	99.7	90.2	72.9	50.4	59.6	28.6
Arikar	55.3	51.6	52.0	66.8	76.6	102.6	79.4	103.9	82.9	76.5	65.4	63.6	70.9	54.8
S.Cruz	49.9	45.7	49.9	56.7	74.0	90.6	70.4	81.6	60.7	68.8	55.0	76.6	90.9	62.4
Peru	46.6	42.8	50.8	55.1	71.0	81.6	66.9	92.0	73.7	75.5	51.0	61.7	69.2	60.4
NJapan	63.2	53.3	63.2	62.0	70.1	70.6	53.7	82.7	89.3	85.2	59.5	59.2	63.0	59.2
SJapan	68.1	57.6	69.9	61.9	63.8	71.1	55.8	85.0	102.7	98.0	67.6	61.3	60.5	65.3
Hainan	61.1	52.3	61.6	60.0	66.0	66.1	58.9	83.9	95.8	88.2	63.8	59.8	59.9	64.7
Anyang	69.7	62.5	76.6	72.5	69.8	80.7	69.1	86.3	106.3	96.1	69.8	57.1	52.3	60.5
Phil	62.5	50.4	57.1	64.0	61.3	65.7	52.2	70.6	85.1	81.7	60.3	69.7	58.5	77.5
Guam	62.4	47.4	58.7	56.4	61.9	60.0	48.0	71.3	80.5	72.6	59.3	56.2	63.5	67.5
Atayal	63.9	52.5	71.4	60.6	62.3	70.3	59.4	91.1	75.7	81.7	55.6	44.6	54.2	63.0
Ainu	50.9	46.8	59.3	51.2	68.7	66.7	55.8	82.8	69.6	79.5	64.4	66.7	64.8	60.6
Andamn	63.4	59.0	64.8	49.7	61.5	57.1	59.2	68.6	85.4	78.2	64.4	76.5	82.3	76.9
Buriat	87.1	80.0	70.0	106.8	107.5	123.7	107.1	110.2	135.9	128.4	113.7	102.7	119.9	105.6
Eskimo	75.3	70.2	85.9	82.2	74.3	117.7	92.0	107.8	90.3	110.1	72.4	88.1	82.6	93.1

	Arikar	S.Cruz	Peru	NJapan	SJapan	Hainan	Anyang	Phil	Guam	Atayal	Ainu	Andamn	Buriat	Eskimo
Norse	55.3	49.9	46.6	63.2	68.1	61.1	69.7	62.5	62.4	63.9	50.9	63.4	87.1	75.3
Zalvar	51.6	45.7	42.8	53.3	57.6	52.3	62.5	50.4	47.4	52.5	46.8	59.0	80.0	70.2
Berg	52.0	49.9	50.8	63.2	69.9	61.6	76.6	57.1	58.7	71.4	59.3	64.8	70.0	85.9
Egypt	66.8	56.7	55.1	62.0	61.9	60.0	72.5	64.0	56.4	60.6	51.2	49.7	106.8	82.2
Teita	76.6	74.0	71.0	70.1	63.8	66.0	69.8	61.3	61.9	62.3	68.7	61.5	107.5	74.3
Dogon	102.6	90.6	81.6	70.6	71.1	66.1	80.7	65.7	60.0	70.3	66.7	57.1	123.7	117.7
Zulu	79.4	70.4	66.9	53.7	55.8	58.9	69.1	52.2	48.0	59.4	55.8	59.2	107.1	92.0
Bush	103.9	81.6	92.0	82.7	85.0	83.9	86.3	70.6	71.3	91.1	82.8	68.6	110.2	107.8
S.Aust	82.9	60.7	73.7	89.3	102.7	95.8	106.3	85.1	80.5	75.7	69.6	85.4	135.9	90.3
Tasman	76.5	68.8	75.5	85.2	98.0	88.2	96.1	81.7	72.6	81.7	79.5	78.2	128.4	110.1
Tolai	65.4	55.0	51.0	59.5	67.6	63.8	69.8	60.3	59.3	55.6	64.4	64.4	113.7	72.4
Mokapu	63.6	76.6	61.7	59.2	61.3	59.8	57.1	69.7	56.2	44.6	66.7	76.5	102.7	88.1
Easter	70.9	90.9	69.2	63.0	60.5	59.9	52.3	58.5	63.5	54.2	64.8	82.3	119.9	82.6
Morior	54.8	62.4	60.4	59.2	65.3	64.7	60.5	77.5	67.5	63.0	60.6	76.9	105.6	93.1
Arikar	25.8	40.9	39.3	57.9	57.1	48.5	59.0	53.9	52.1	53.6	66.3	61.5	73.3	76.2
S.Cruz	40.9	23.8	35.4	50.9	59.3	55.9	68.2	56.9	54.7	59.9	60.2	52.3	88.1	73.1
Peru	39.3	35.4	23.4	47.2	51.4	42.4	54.0	45.0	47.8	48.9	60.5	54.7	88.9	77.6
NJapan	57.9	50.9	47.2	27.9	30.8	35.9	38.7	39.9	37.6	45.8	46.4	59.9	78.4	70.8
SJapan	57.1	59.3	51.4	30.8	26.0	31.3	33.9	37.0	35.5	42.4	53.1	54.6	80.7	68.0
Hainan	48.5	55.9	42.4	35.9	31.3	26.8	31.8	32.0	31.2	38.2	49.8	49.0	76.4	76.7
Anyang	59.0	68.2	54.0	38.7	33.9	31.8	26.8	39.8	37.0	39.5	53.5	64.1	81.5	75.7
Phil	53.9	56.9	45.0	39.9	37.0	32.0	39.8	28.1	36.5	46.6	49.7	50.2	87.9	77.4
Guam	52.1	54.7	47.8	37.6	35.5	31.2	37.0	36.5	25.9	36.2	49.4	46.8	78.4	84.2
Atayal	53.6	59.9	48.9	45.8	42.4	38.2	39.5	46.6	36.2	23.9	47.4	60.9	79.6	59.0
Ainu	66.3	60.2	60.5	46.4	53.1	49.8	53.5	49.7	49.4	47.4	29.2	64.6	93.1	71.3
Andamn	61.5	52.3	54.7	59.9	54.6	49.0	64.1	50.2	46.8	60.9	64.6	24.1	112.9	98.0
Buriat	73.3	88.1	88.9	78.4	80.7	76.4	81.5	87.9	78.4	79.6	93.1	112.9	33.7	87.7
Eskimo	76.2	73.1	77.6	70.8	68.0	76.7	75.7	77.4	84.2	59.0	71.3	98.0	87.7	31.1

TABLE 9A-I
POPKIN results: Peru TEST #1426

```
Peru    1    2    3    4    5    7    8   10   11   12   13   14   19   20   21     22   24   25   28   31   34   38   45   47   51   69  103  139  151  202     39
        3.5  3.7  3.8  3.8  4.0  4.1  4.2  4.2  4.3  4.4  4.4  4.5  4.6  4.7  4.7    4.8  4.8  4.9  5.0  5.0  5.1  5.2  5.3  5.3  5.3  5.6  5.9  6.2  6.2  6.5    4.8

Zalava  15   30   39   48   49   54   68   72   77   80   87   88   92   97   99    102  120  121  129  131  136  153  178  186  195  201  251  265  300  484    132
        4.5  5.0  5.2  5.3  5.3  5.3  5.6  5.6  5.7  5.7  5.8  5.8  5.8  5.8  5.9    5.9  6.0  6.0  6.1  6.1  6.2  6.2  6.4  6.5  6.5  6.5  6.8  6.9  7.0  7.8    6.0

S.Cruz  17   33   37   53   57   58   60   79   81   90  101  116  125  132  140    179  183  188  200  206  218  264  270  285  288  293  317  368  375  394    174
        4.6  5.1  5.2  5.3  5.4  5.4  5.5  5.7  5.7  5.8  5.9  6.0  6.1  6.1  6.2    6.4  6.4  6.5  6.5  6.6  6.6  6.9  6.9  6.9  6.9  7.0  7.1  7.3  7.3  7.4    6.2

Atayal  6    23   26   35   44   52   55   65   94   98  110  118  135  137  142    143  147  187  191  207  214  247  275  309  388  452  564  606  650    .    197
        4.0  4.8  4.9  5.1  5.2  5.3  5.4  5.6  5.8  5.9  5.9  6.0  6.1  6.2  6.2    6.2  6.2  6.5  6.5  6.6  6.6  6.8  6.9  7.0  7.3  7.6  8.2  8.4  8.7    .    6.3

Berg    43   50   59   61   74   78   85  104  108  113  134  138  148  157  169    248  259  262  321  334  340  358  367  386  389  399  405  492  513  520    237
        5.2  5.3  5.5  5.6  5.7  5.7  5.7  5.9  5.9  6.0  6.1  6.2  6.2  6.3  6.4    6.8  6.8  6.9  7.1  7.1  7.2  7.2  7.3  7.3  7.3  7.4  7.4  7.8  7.9  7.9    6.6

Hainan  9    18   66   73  109  126  128  160  192  199  204  213  233  240  241    250  253  257  266  286  298  302  313  352  385  398  425  435  457  620    250
        4.2  4.6  5.6  5.6  5.9  6.1  6.1  6.3  6.5  6.5  6.5  6.6  6.7  6.7  6.7    6.8  6.8  6.8  6.9  6.9  7.0  7.0  7.1  7.2  7.3  7.4  7.5  7.5  7.6  8.5    6.6

Guam    27   29   41   91  106  107  122  162  167  181  190  220  228  268  282    296  301  306  308  343  371  372  376  381  390  393  466  468  512  749    275
        5.0  5.0  5.2  5.8  5.9  5.9  6.0  6.3  6.4  6.4  6.5  6.7  6.7  6.9  6.9    7.0  7.0  7.0  7.0  7.2  7.3  7.3  7.3  7.3  7.3  7.3  7.7  7.7  7.9  9.3    6.8

Norse   32   36   42   63   93  100  145  149  163  165  166  217  223  226  231    280  297  322  327  350  356  365  408  437  447  465  488  598  645  660    283
        5.1  5.2  5.2  5.6  5.8  5.9  6.2  6.2  6.3  6.3  6.3  6.6  6.7  6.7  6.7    6.9  7.0  7.1  7.1  7.2  7.2  7.3  7.4  7.5  7.6  7.7  7.8  8.4  8.6  8.7    6.8

Arikar  16   46   62   67   89  105  114  194  197  203  221  232  258  272  276    281  283  361  406  413  417  471  476  533  547  553  573  599  613  747    321
        4.5  5.3  5.6  5.6  5.8  5.9  6.0  6.5  6.5  6.5  6.7  6.7  6.8  6.9  6.9    6.9  6.9  7.2  7.4  7.4  7.4  7.7  7.7  8.0  8.1  8.1  8.3  8.4  8.4  9.3    7.0

Philip  56   83  111  124  152  180  205  219  222  229  230  244  274  284  332    349  355  395  404  432  434  440  444  455  462  481  510  525  538  618    326
        5.4  5.7  6.0  6.0  6.2  6.4  6.6  6.7  6.7  6.7  6.7  6.8  6.9  6.9  7.1    7.2  7.2  7.4  7.4  7.5  7.5  7.6  7.6  7.6  7.7  7.8  7.9  8.0  8.0  8.5    7.1

NJapan  64   70  130  141  176  184  212  216  239  255  260  263  271  273  312    318  335  348  384  392  433  456  477  489  503  508  539  552  601  781    339
        5.6  5.6  6.1  6.2  6.4  6.4  6.6  6.6  6.7  6.8  6.8  6.9  6.9  6.9  7.1    7.1  7.1  7.2  7.3  7.3  7.5  7.6  7.7  7.8  7.8  7.9  8.0  8.1  8.4  9.7    7.1

Tolai   76   96  119  146  154  158  164  174  189  211  235  238  245  256  278    287  294  363  369  373  391  441  571  572  587  597  604  667  694  705    342
        5.7  5.8  6.0  6.2  6.2  6.3  6.3  6.4  6.5  6.6  6.7  6.7  6.8  6.8  6.9    6.9  7.0  7.3  7.3  7.3  7.3  7.6  8.3  8.3  8.3  8.4  8.4  8.8  8.9  9.0    7.2

SJapan  112  115  127  133  161  209  225  310  314  315  328  329  357  359  364    374  382  383  402  403  420  467  469  479  527  575  581  583  628  672    370
        6.0  6.0  6.1  6.1  6.3  6.6  6.7  7.1  7.1  7.1  7.1  7.1  7.2  7.2  7.3    7.3  7.3  7.3  7.4  7.4  7.4  7.7  7.7  7.7  7.8  8.0  8.3  8.3  8.3  8.5  8.8    7.3

Egypt   68   71   75  123  144  175  177  185  240  320  378  387  422  430  448    443  453  466  463  508  ...                                          617
        5.2  5.6  5.7  6.0  6.2  6.4  6.4  6.4  7.0  7.1  7.3  7.3  7.4  7.5  7.6    7.6  7.6  7.7  7.8  7.9  7.9  7.9  8.2  8.5  8.5  8.5  8.9  9.9  10   10    7.5

Anyang  82   86  155  156  168  171  210  289  291  311  360  377  423  460  482    497  505  518  524  529  561  579  591  593  605  657  669  692  751  756    435
        5.7  5.8  6.2  6.2  6.4  6.4  6.6  7.0  7.0  7.1  7.2  7.3  7.4  7.6  7.8    7.8  7.9  7.9  8.0  8.0  8.2  8.3  8.4  8.4  8.4  8.7  8.8  8.9  9.4  9.5    7.6

Andama  182  196  243  249  252  254  304  307  333  344  345  366  379  414  421    442  450  454  486  494  536  542  577  590  627  643  656  676  702  724    440
        6.4  6.5  6.8  6.8  6.8  6.8  6.7  7.0  7.0  7.1  7.2  7.2  7.3  7.3  7.4    7.6  7.6  7.6  7.8  7.8  8.0  8.1  8.3  8.4  8.5  8.6  8.7  8.8  9.0  9.2    7.6

Eskimo  159  173  208  267  269  277  279  316  336  337  338  339  370  410  427    439  458  483  487  586  608  623  625  641  663  670  711  716  791  832    461
        6.3  6.4  6.6  6.9  6.9  6.9  6.9  7.1  7.1  7.1  7.2  7.2  7.3  7.4  7.5    7.6  7.6  7.8  7.8  8.3  8.4  8.5  8.5  8.6  8.7  8.8  8.9  9.1  9.1  9.8  11    7.8

Zulu    150  198  227  234  237  342  346  362  396  424  429  443  472  478  491    568  570  634  647  654  664  678  683  698  704  719  733  759  767  788    517
        6.2  6.5  6.7  6.7  6.7  7.2  7.2  7.2  7.4  7.5  7.5  7.6  7.7  7.7  7.8    8.2  8.2  8.5  8.6  8.7  8.8  8.8  8.9  9.0  9.0  9.1  9.2  9.5  9.6  9.8    8.1

Ainu    117  170  193  261  292  401  446  475  500  502  504  511  517  523  537    540  566  578  582  611  619  659  695  696  697  701  715  728  738  740    527
        6.0  6.4  6.5  6.8  7.0  7.4  7.6  7.7  7.8  7.8  7.9  7.9  7.9  7.9  8.0    8.0  8.2  8.3  8.3  8.4  8.5  8.7  8.9  9.0  9.0  9.0  9.1  9.2  9.3  9.3    8.1

Teita   224  236  323  325  341  347  397  400  419  438  490  493  496  521  534    544  548  551  554  556  602  610  661  675  707  713  753  777  795  820    528
        6.7  6.7  7.1  7.1  7.2  7.2  7.4  7.4  7.4  7.6  7.8  7.8  7.8  7.9  8.0    8.1  8.1  8.1  8.1  8.1  8.4  8.4  8.7  8.8  9.0  9.1  9.4  9.6  9.9  10    8.1

Austra  95   324  354  409  416  418  436  451  463  495  499  515  516  528  535    545  607  636  651  703  709  717  726  735  744  757  775  792  827  829    574
        5.8  7.1  7.2  7.4  7.4  7.4  7.5  7.6  7.7  7.8  7.8  7.9  7.9  8.0  8.0    8.1  8.4  8.5  8.7  9.0  9.1  9.2  9.2  9.3  9.5  9.6  9.9  11   11    8.4

Hawaii  172  246  380  448  453  464  473  522  532  550  558  562  589  592  646    648  649  655  658  665  673  680  681  700  712  766  778  786  805  825    596
        6.4  6.8  7.3  7.6  7.6  7.7  7.7  7.9  8.0  8.1  8.2  8.2  8.4  8.4  8.6    8.6  8.6  8.7  8.7  8.8  8.8  8.8  8.9  9.0  9.1  9.5  9.7  9.8  10   11    8.5

Bushma  330  428  461  501  507  509  530  541  543  563  567  609  612  614  616    622  626  630  638  640  652  668  720  741  743  802  806  811  812  837    623
        7.1  7.5  7.6  7.8  7.9  7.9  8.0  8.1  8.1  8.2  8.2  8.4  8.4  8.4  8.4    8.5  8.5  8.5  8.6  8.6  8.7  8.9  9.1  9.3  9.3  10   10   10   10   12    8.7

Buriat  84   215  303  407  415  470  569  574  594  603  615  637  689  691  699    718  727  731  739  745  752  755  772  773  784  809  822  835  836  839    640
        5.7  6.6  7.0  7.4  7.4  7.8  8.2  8.3  8.4  8.4  8.4  8.5  8.9  8.9  9.0    9.1  9.2  9.2  9.3  9.3  9.4  9.4  9.6  9.6  9.8  10   10   11   11   12    8.9

Easter  412  426  474  526  531  549  555  557  588  595  596  600  621  629  631    642  653  662  666  710  725  748  754  768  769  771  774  785  790  813    644
        7.4  7.5  7.7  8.0  8.0  8.1  8.1  8.2  8.3  8.4  8.4  8.4  8.5  8.5  8.5    8.6  8.7  8.7  8.8  9.1  9.2  9.3  9.4  9.6  9.6  9.6  9.6  9.8  9.8  10    8.7

Tasman  242  299  326  351  546  617  633  639  671  674  679  682  684  685  690    693  721  722  729  750  758  761  770  779  797  798  808  826  828  833    666
        6.8  7.0  7.1  7.2  8.1  8.5  8.5  8.6  8.8  8.8  8.8  8.8  8.9  8.9  9.1    8.9  9.1  9.2  9.2  9.3  9.5  9.5  9.6  9.7  9.9  9.9  10   11   11   11    9.0

Dogon   295  319  331  353  498  559  580  585  624  632  644  688  730  732  734    762  765  776  782  783  789  794  796  801  803  814  818  821  823  824    675
        7.0  7.1  7.1  7.2  7.8  8.2  8.3  8.3  8.5  8.5  8.6  8.9  9.2  9.2  9.2    9.5  9.5  9.6  9.7  9.7  9.8  9.9  9.9  10   10   10   10   10   11    9.1

Morior  305  411  431  560  677  687  706  708  714  723  736  737  742  746  763    764  780  787  793  799  800  807  810  815  817  819  830  831  834  838    726
        7.0  7.4  7.5  8.2  8.8  8.9  9.0  9.0  9.1  9.2  9.2  9.3  9.3  9.3  9.5    9.5  9.7  9.8  9.9  9.9  10   10   10   10   10   10   11   11   11   12    9.5
```

median distance at the center of the line. Note also that this mean distance, for a single skull, is not the same as the distances in DISPOP, which are those of a target skull to a population *mean*.

With 839 control skulls, the median distance rank is 420, which thus divides the whole body into nearer and more distant moieties of the whole control set. In the case of #1426, the rankings for the Peru control sample not only fall entirely within the nearer moiety (i.e., with rankings less than 420) but in fact within the lowest quarter of rankings—that is, the most distant Peruvian control is 202 in rank.

In other figures this apparently "average" Peruvian skull positions itself clearly in the Peruvian part of the reference space. The mean sample distance in the Peru line is 4.8 and mean rank 39, far below other populations. In the nearer half of the line (i.e., in the nearest 15) there is no ranking higher than 21 and only 5 of the 30 Peru controls have a distance larger than 5.3 (i.e., the mean of intragroup distances—see page 50.)

The next population in proximity is Zalavár, well behind Peru. The nearest control is at rank 15, the mean distance is 6.0, and there are only 6 controls at a distance of 5.3 or less. Atayal is the next nearest, with some lower rankings early but higher ranks in the second fifteen.

It is easy to exclude populations from consideration. Bush, for example, has only one control in the lower moiety of rankings and Buriats have 5, but also 3 of the 5 most distant controls, with ranks of 835, 836, and 839. Also obviously excludable are Africans, Australo-Melanesians and Polynesians. Least distant are other Americans, Europeans and some Asians.

Peruvian TEST #1414 (Table 9A-2) seems supertypical for that population with, in DISPOP, a very low D of 1.8 and a P of .99, close to the central point of the Peru space. POPKIN elaborates; the most distant Peru control is of rank 124, and this is the only one of the Peru controls with a distance from #1414 greater than 5.3. At the same time, this specimen is not far from the spaces of the other two Indian groups. Santa Cruz has the nearest control of all, with some early low ranks, while Arikara is generally lower ranked over much of the middle of the range. Again, this median point is a good place to look for ranking comparisons among control populations. Once more, obviously excludable are Africans, Australo-Melanesians, and Polynesians; Europeans and Asiatics are not suggestively close.

Peruvian TEST #1435 (table 9A-3) was emphatically European in the DISPOP diagnosis, and is so here, with Peru being elbowed out of the first 39 positions and showing only 6 controls with a distance of 5.3 or less. Overall, Peru comes in as fifth choice. While Americans are, in general, next nearest compared to Europeans, they are not close and assignment of this skull would have to be European, preferably southern. Every way of reading this table comes to that result.

Here I would heed my admonition expressed earlier, that such departures and wrong assignments are to be expected in a small proportion of cases. I was eventually so impressed by the emphatic European placement of #1435 that I went to the collection (fortunately in the Peabody Museum) to have another look at this skull. It is indeed European in appearance, having a prominent nasal root, a narrow high nose, well-developed supraorbitals for its general degree of robustness, and only moderately wide malars. The effect of supraorbitals in producing a "European" affiliation, though difficult to find in the computations, may be important.

Maori tests

When POPKIN is applied to the 20 Maori skulls, results seem less "good" than those found by DISPOP above, though more fully informative. With DISPOP, three of twenty specimens are read as non-Polynesian (Anyang, Arikara, Ainu). With POPKIN, six (including the same three) appear as non-Polynesian, by closest placement of the mean distance (last column) and, in general, by the whole array of members of a control sample: POPKIN diagnoses these six as 2 Anyang, 2 Arikara, 1 Ainu, and 1 Tolai.

Table 9B displays three of the Maori cases. These three cases do not include those most positively identified with a Polynesian group, but rather illustrate some other points.

Number 3029 (Table 9B-1) has, by DISPOP, these probabilities: .44 of being Easter, .28 of being Hawaiian, and .22 of being Guam. In the

Who's Who in Skulls

POPKIN results: Peru TEST #1414

```
Peru      2    3    4    5    6    8    9   10   11   12   15   17   19   20   22    34   40   41   42   43   44   46   49   54   55   56   77   87   89  124    35
        3.0  3.1  3.1  3.2  3.2  3.3  3.3  3.4  3.4  3.5  3.6  3.8  3.9  4.0  4.1   4.3  4.4  4.4  4.4  4.4  4.4  4.5  4.5  4.7  4.7  4.7  5.1  5.2  5.2  5.5   4.1

Arikar   13   14   18   25   28   29   30   33   35   37   39   53   59   69   73    82  104  106  108  110  127  128  148  167  190  212  238  276  303  340   106
        3.5  3.6  3.8  4.1  4.1  4.1  4.2  4.3  4.3  4.4  4.4  4.6  4.7  4.9  5.0   5.1  5.3  5.4  5.4  5.4  5.5  5.6  5.7  5.8  6.0  6.1 -6.2  6.4  6.6  6.8   5.0

S.Cruz    1    7   24   26   27   31   36   38   47   57   64   66   70   81   93   115  129  132  133  142  145  164  165  179  185  205  208  292  314  329   117
        2.9  3.3  4.1  4.1  4.1  4.2  4.4  4.4  4.5  4.7  4.9  4.9  5.0  5.1  5.3   5.5  5.5  5.6  5.6  5.7  5.7  5.8  5.8  5.9  5.9  6.0  6.0  6.5  6.6  6.7   5.2

Hainan   21   32   48   51   75   90  100  101  113  117  118  143  150  156  170   175  192  193  204  209  226  240  271  294  298  312  350  376  383  406   187
        4.0  4.2  4.5  4.6  5.0  5.2  5.3  5.3  5.5  5.5  5.5  5.5  5.7  5.7  5.7   5.9  6.0  6.0  6.0  6.0  6.1  6.2  6.4  6.5  6.5  6.6  6.8  6.9  7.0  7.1   5.8

Atayal   16   62   65   71   72   86  105  114  134  151  172  177  178  194  196   231  247  249  297  302  307  336  343  378  380  390  396  443  457    .   226
        3.8  4.9  4.9  5.0  5.0  5.2  5.4  5.5  5.6  5.7  5.8  5.9  5.9  6.0  6.0   6.2  6.3  6.3  6.5  6.6  6.6  6.7  6.8  6.9  6.9  7.0  7.0  7.2  7.3    .   6.0

Zalava   60   74   94  109  120  123  158  161  162  173  198  199  200  207  228   242  263  283  310  317  333  375  399  440  445  449  453  523  527  614   275
        4.8  5.0  5.3  5.4  5.5  5.5  5.8  5.8  5.8  5.8  6.0  6.0  6.0  6.0  6.2   6.2  6.3  6.4  6.6  6.6  6.7  6.9  7.0  7.2  7.2  7.2  7.3  7.6  7.6  8.1   6.3

Andama   85  112  131  135  137  138  147  169  171  189  197  201  202  219  230   233  248  250  264  285  323  328  398  418  421  428  432  582  605  640   277
        5.2  5.5  5.6  5.6  5.7  5.7  5.7  5.8  5.8  6.0  6.0  6.0  6.0  6.1  6.2   6.2  6.3  6.3  6.3  6.5  6.7  6.7  7.0  7.2  7.2  7.2  7.2  7.9  8.0  8.3   6.4

Norse    45   63   67  116  119  122  136  155  163  168  210  216  239  244  259   277  287  305  335  349  361  369  422  435  439  451  458  537  572  615   284
        4.4  4.9  4.9  5.5  5.5  5.5  5.6  5.7  5.8  5.8  6.0  6.1  6.2  6.2  6.3   6.4  6.5  6.6  6.7  6.8  6.9  6.9  7.2  7.2  7.2  7.3  7.3  7.7  7.8  8.1   6.4

Philip   78   84   98  102  103  121  180  187  191  235  243  258  269  272  280   289  304  313  337  362  385  397  408  417  466  524  534  542  551  593   304
        5.1  5.2  5.3  5.3  5.3  5.5  5.9  5.9  6.0  6.2  6.2  6.3  6.4  6.4  6.4   6.5  6.6  6.6  6.7  6.9  7.0  7.0  7.1  7.2  7.3  7.6  7.6  7.7  7.8  8.0   6.5

Berg     52   95   96  126  139  141  160  181  183  221  232  251  257  295  300   316  346  357  373  389  412  423  424  463  483  540  573  599  613  693   324
        4.6  5.3  5.3  5.5  5.7  5.7  5.8  5.9  5.9  6.1  6.2  6.3  6.3  6.5  6.5   6.6  6.8  6.8  6.9  7.0  7.1  7.2  7.2  7.3  7.4  7.7  7.8  8.0  8.1  8.7   6.6

NJapan   61   92  111  125  154  157  188  203  218  222  246  252  299  324  331   356  374  379  391  403  411  431  447  467  468  516  529  538  587  718   333
        4.8  5.3  5.5  5.5  5.7  5.8  6.0  6.0  6.1  6.1  6.3  6.3  6.5  6.7  6.7   6.8  6.9  6.9  7.0  7.0  7.1  7.2  7.2  7.3  7.3  7.5  7.6  7.7  7.9  8.8   6.7

SJapan   76   91   97  140  146  217  227  254  260  267  279  318  319  347  360   366  367  370  377  416  436  454  464  474  490  505  549  584  620  669   351
        5.0  5.2  5.3  5.7  5.7  6.1  6.1  6.3  6.3  6.4  6.4  6.6  6.6  6.8  6.9   6.9  6.9  6.9  7.1  7.2  7.3  7.3  7.4  7.4  7.5  7.7  7.9  8.2  8.5   6.7

Guam     80   83  159  195  214  229  253  266  268  270  273  275  309  334  353   382  394  407  413  420  427  438  441  469  486  509  563  569  586  733   360
        5.1  5.2  5.8  6.0  6.1  6.2  6.3  6.4  6.4  6.4  6.4  6.4  6.6  6.7  6.8   7.0  7.0  7.1  7.1  7.2  7.2  7.2  7.2  7.3  7.4  7.5  7.0  7.0  7.9  9.0   6.8

Egypt    68   88   99  107  144  182  186  206  215  220  224  255  262  311  352   354  386  435  444  462  471  500  506  531  536  562  611  700  726  770   364
        4.9  5.2  5.3  5.4  5.7  5.9  5.9  6.0  6.1  6.1  6.1  6.3  6.3  6.6  6.8   6.8  7.0  7.2  7.2  7.3  7.3  7.5  7.5  7.6  7.6  7.8  8.1  8.7  8.9  9.4   6.8

Anyang   23   79  234  236  245  265  281  286  326  330  339  372  409  446  473   480  489  499  528  541  553  556  559  560  567  570  622  677  692  716   432
        4.1  5.1  6.2  6.2  6.3  6.4  6.4  6.5  6.7  6.7  6.8  6.9  7.1  7.2  7.4   7.4  7.4  7.5  7.6  7.7  7.7  7.7  7.8  7.8  7.8  7.8  8.2  8.5  8.6  8.8   7.1

Ainu    149  184  274  284  320  327  341  342  348  355  371  381  402  405  437   448  452  470  481  485  498  545  550  579  618  630  633  638  717  729   446
        5.7  5.9  6.4  6.4  6.7  6.7  6.8  6.8  6.8  6.8  6.9  6.9  7.0  7.1  7.2   7.2  7.3  7.3  7.4  7.4  7.5  7.7  7.7  7.9  8.1  8.2  8.3  8.3  8.9  9.0   7.3

Morior   50   58  153  290  291  296  301  306  338  351  359  392  395  410  426   513  555  558  564  566  568  583  585  596  601  612  617  652  727  738   448
        4.6  4.7  5.7  6.5  6.5  6.5  6.6  6.6  6.8  6.8  6.8  6.8  7.0  7.1  7.2   7.5  7.7  7.8  7.8  7.8  7.9  7.9  8.0  8.0  8.1  8.1  8.4  8.9  9.0   7.2

Tolai   152  174  225  237  241  261  293  345  365  368  460  472  482  487  492   493  494  495  497  507  508  517  557  636  666  674  707  712  719  809   468
        5.7  5.9  6.1  6.2  6.2  6.3  6.5  6.8  6.9  6.9  7.3  7.3  7.4  7.4  7.4   7.5  7.5  7.5  7.5  7.5  7.5  7.6  7.7  8.3  8.5  8.5  8.7  8.8  8.8   10   7.4

Zulu    166  308  315  322  332  364  425  478  503  520  522  526  535  539  548   595  598  600  608  632  634  642  643  644  655  661  676  720  751  815   543
        5.8  6.6  6.6  6.7  6.7  6.9  7.2  7.4  7.5  7.6  7.6  7.6  7.6  7.7  7.7   8.0  8.0  8.0  8.1  8.2  8.3  8.3  8.3  8.3  8.4  8.4  8.5  8.9  9.2   10   7.8

Hawaii  213  256  288  401  414  415  429  455  456  512  519  525  571  575  580   590  604  610  628  687  695  703  704  723  746  755  763  764  777  800   572
        6.1  6.3  6.5  7.0  7.1  7.1  7.2  7.3  7.3  7.5  7.6  7.6  7.8  7.9  7.9   7.9  8.0  8.1  8.2  8.6  8.7  8.7  8.7  8.9  9.2  9.3  9.4  9.4  9.5  9.9   8.0

Teita   223  325  387  388  430  484  488  502  504  510  514  518  532  544  591   594  602  616  619  629  637  684  690  694  731  749  752  754  771  832   576
        6.1  6.7  7.0  7.0  7.2  7.4  7.4  7.5  7.5  7.5  7.5  7.6  7.6  7.8  8.0   8.0  8.0  8.1  8.1  8.2  8.3  8.5  8.6  8.7  9.0  9.2  9.2  9.3  9.4   11   8.0

Dogon   278  321  384  393  434  515  547  561  581  589  609  621  635  645  670   671  672  682  689  698  701  721  741  743  748  765  767  775  778  830   625
        6.4  6.7  7.0  7.0  7.2  7.5  7.7  7.8  7.9  7.9  8.1  8.2  8.3  8.3  8.5   8.5  8.5  8.5  8.6  8.7  8.7  8.9  9.1  9.1  9.2  9.4  9.4  9.5  9.5   10   8.3

Austra  176  363  404  461  475  477  511  554  574  576  578  592  597  623  631   640  647  654  667  705  757  758  761  766  772  773  802  817  823  829   626
        5.9  6.9  7.1  7.3  7.4  7.4  7.5  7.7  7.8  7.9  7.9  8.0  8.0  8.2  8.2   8.3  8.3  8.4  8.5  8.7  9.3  9.3  9.3  9.4  9.4  9.4  9.9   10   10   10   8.4

Easter  344  442  476  479  491  533  546  552  577  607  624  627  648  651  656   660  697  699  702  722  730  732  734  735  736  750  792  806  808  819   646
        6.8  7.2  7.4  7.4  7.4  7.6  7.7  7.7  7.9  8.1  8.2  8.2  8.3  8.4  8.4   8.4  8.7  8.7  8.7  8.9  9.0  9.0  9.0  9.0  9.0  9.0  9.2  9.7   10   10   10   8.5

Tasman  211  282  358  400  450  588  625  649  653  662  663  683  696  708  709   715  739  756  759  762  774  781  788  793  796  803  811  814  828  839   670
        6.1  6.4  6.8  7.0  7.3  7.9  8.2  8.4  8.4  8.5  8.5  8.5  8.7  8.7  8.8   8.9  9.1  9.3  9.3  9.3  9.4  9.6  9.6  9.8  9.8  9.9   10   10   10   12   8.8

Eskimo  419  459  496  530  543  565  603  641  665  668  678  685  686  688  725   740  744  753  760  768  769  776  783  795  798  804  805  816  821  834   694
        7.2  7.3  7.5  7.6  7.7  7.8  8.0  8.3  8.5  8.5  8.5  8.5  8.5  8.6  8.9   9.1  9.2  9.2  9.3  9.4  9.4  9.5  9.6  9.8  9.8  9.9  9.9   10   10   11   8.9

Buriat  130  465  501  521  606  626  639  659  664  679  691  711  713  714  728   745  747  780  782  791  794  799  801  820  822  825  827  831  835  836   703
        5.6  7.3  7.5  7.6  8.0  8.2  8.3  8.4  8.5  8.5  8.6  8.8  8.8  8.8  9.0   9.2  9.2  9.6  9.6  9.7  9.8  9.9  9.9   10   10   10   10   10   11   12   9.1

Bushma  650  657  658  673  675  680  681  706  710  724  737  742  779  784  785   786  787  789  790  797  807  810  812  813  818  824  826  833  837  838   760
        8.4  8.4  8.4  8.5  8.5  8.5  8.5  8.7  8.8  8.9  9.0  9.1  9.5  9.6  9.6   9.6  9.6  9.7  9.7  9.8   10   10   10   10   10   10   10   11   12   12   9.6
```

TABLE 9A-3
POPKIN results: Peru TEST #1435

```
Berg     1    3    5   10   11   12   17   19   21   25   26   30   49   53   60    74   77   84   86   95  102  103  113  118  121  149  185  204  218  316    80
       2.5  3.7  3.7  4.1  4.2  4.2  4.4  4.5  4.5  4.6  4.6  4.7  4.9  4.9  5.1   5.2  5.3  5.3  5.3  5.4  5.5  5.5  5.6  5.6  5.7  5.8  6.1  6.2  6.3  6.9   5.0

Zalava   2    6    9   13   15   22   24   28   29   32   33   38   50   51   62    79   88   93  105  107  120  123  124  127  163  182  190  247  322  394    96
       3.2  3.7  4.0  4.2  4.3  4.5  4.6  4.6  4.7  4.7  4.7  4.8  4.9  4.9  5.1   5.3  5.3  5.4  5.5  5.5  5.6  5.7  5.7  5.7  5.9  6.1  6.1  6.5  6.9  7.3   5.2

Norse    4   14   31   35   41   46   48   61   68   69   72   90   92   96  100   106  111  112  137  161  168  177  196  221  224  246  280  304  624  711   155
       3.7  4.3  4.7  4.7  4.8  4.9  4.9  5.1  5.2  5.2  5.2  5.4  5.4  5.4  5.5   5.5  5.6  5.6  5.8  5.9  5.9  6.0  6.2  6.3  6.3  6.5  6.7  6.8  8.3  8.9   5.7

Andama  34   39   42   44   57   58   63   76   82   89   99  108  109  128  145   155  160  200  220  233  235  274  275  277  291  296  344  359  371  396   175
       4.7  4.8  4.8  4.9  5.1  5.1  5.2  5.2  5.3  5.4  5.5  5.5  5.5  5.5  5.7   5.8  5.9  6.2  6.3  6.4  6.4  6.7  6.7  6.7  6.7  6.7  7.0  7.1  7.1  7.3   5.9

Egypt    8   16   20   36   71   81   87  110  122  125  130  147  152  158  173   178  189  201  202  208  210  255  290  311  332  337  362  492  498  582   203
       3.9  4.3  4.5  4.7  5.2  5.3  5.3  5.6  5.7  5.7  5.7  5.8  5.8  5.9  6.0   6.0  6.1  6.2  6.2  6.3  6.3  6.6  6.7  6.8  6.9  7.0  7.1  7.7  7.8  8.2   6.0

Peru    40   43   52   64   67   85   98  114  132  144  151  154  172  174  195   197  198  203  205  228  230  234  256  293  326  366  405  417  427  681   212
       4.8  4.9  4.9  5.2  5.2  5.3  5.5  5.6  5.7  5.8  5.8  6.0  6.0  6.1  6.1   6.2  6.2  6.2  6.2  6.4  6.4  6.4  6.6  6.6  6.7  6.9  7.1  7.3  7.4  7.4  8.6  6.2

Arikar  18   23   47   91  117  126  134  146  157  167  181  191  194  226  237   248  258  278  281  306  327  338  346  363  395  415  425  435  505  680   255
       4.4  4.6  4.9  5.4  5.6  5.7  5.8  5.8  5.8  5.9  6.1  6.1  6.1  6.4  6.4   6.5  6.6  6.7  6.7  6.8  6.9  7.0  7.0  7.1  7.3  7.3  7.4  7.4  7.8  8.6   6.4

Atayal  55   56   65   66   73   78  138  139  142  150  153  165  170  244  249   250  260  268  284  303  335  382  421  428  437  596  611  618  717    .   271
       5.0  5.1  5.2  5.2  5.2  5.3  5.8  5.8  5.8  5.8  5.8  5.9  6.0  6.5  6.5   6.5  6.6  6.6  6.7  6.8  6.9  7.2  7.4  7.4  7.5  8.2  8.3  8.3  8.9    .   6.5

S.Cruz  45   54   59   83  115  129  156  171  176  187  188  219  245  262  271   287  318  323  330  336  347  365  390  449  473  475  509  544  557  594   289
       4.9  5.0  5.1  5.3  5.6  5.7  5.8  6.0  6.0  6.1  6.1  6.3  6.5  6.6  6.6   6.7  6.9  6.9  6.9  7.0  7.0  7.1  7.2  7.5  7.6  7.7  7.8  8.0  8.0  8.2   6.6

Hainan  70   75  101  104  116  131  136  162  179  207  225  231  253  259  269   297  353  356  368  373  378  387  432  441  451  508  519  577  588  623   306
       5.2  5.2  5.5  5.5  5.6  5.7  5.8  5.9  6.1  6.3  6.3  6.4  6.5  6.6  6.6   6.7  7.0  7.0  7.1  7.1  7.2  7.2  7.4  7.5  7.5  7.8  7.8  8.1  8.2  8.3   6.7

Philip  27   37  119  133  159  183  214  236  242  251  254  265  270  283  286   288  292  300  301  320  329  341  367  379  393  476  574  595  637  656   307
       4.6  4.8  5.6  5.8  5.9  6.1  6.3  6.4  6.5  6.5  6.6  6.6  6.6  6.7  6.7   6.7  6.7  6.7  6.8  6.9  6.9  7.0  7.1  7.2  7.2  7.7  8.1  8.2  8.4  8.5   6.7

SJapan 135  186  217  222  252  307  313  315  317  351  369  374  389  418  438   442  445  467  495  502  512  516  529  536  540  570  602  652  673  708   426
       5.8  6.1  6.3  6.3  6.5  6.8  6.8  6.8  6.9  7.0  7.1  7.2  7.2  7.4  7.5   7.5  7.5  7.6  7.7  7.8  7.8  7.8  7.9  7.9  7.9  8.1  8.3  8.5  8.6  8.9   7.4

NJapan 140  164  212  215  282  325  339  340  345  376  392  413  444  457  461   465  470  500  553  567  576  599  606  609  616  643  668  675  700  756   464
       5.8  5.9  6.3  6.3  6.7  6.9  7.0  7.0  7.0  7.2  7.2  7.3  7.5  7.6  7.6   7.6  7.6  7.8  8.0  8.1  8.1  8.2  8.3  8.3  8.3  8.4  8.5  8.6  8.8  9.2   7.6

Ainu    94  175  241  267  312  319  321  334  383  386  407  411  430  450  483   496  499  526  545  560  564  578  600  639  646  672  679  687  689  821   474
       5.4  6.0  6.5  6.6  6.8  6.9  6.9  6.9  7.2  7.2  7.3  7.3  7.4  7.5  7.7   7.8  7.8  7.9  8.0  8.1  8.1  8.1  8.2  8.4  8.4  8.6  8.6  8.6  8.7   10   7.6

Guam     7  229  261  263  289  349  360  401  408  410  436  458  463  482  485   486  503  510  527  543  563  586  590  598  615  628  670  674  716  764   476
       3.9  6.4  6.6  6.6  6.7  7.0  7.1  7.3  7.3  7.3  7.4  7.6  7.6  7.7  7.7   7.7  7.8  7.8  7.9  7.9  8.1  8.2  8.2  8.2  8.3  8.3  8.6  8.6  8.9  9.3   7.6

Tolai  192  193  216  227  238  257  264  285  333  380  388  416  423  431  439   466  469  477  478  504  603  651  685  693  732  740  755  796  811  832   479
       6.1  6.1  6.3  6.4  6.5  6.6  6.6  6.7  6.9  7.2  7.2  7.3  7.4  7.4  7.5   7.6  7.6  7.7  7.7  7.8  8.3  8.4  8.6  8.8  8.9  9.1  9.2  9.6  9.9   11   7.7

Anyang  97  143  272  295  342  343  348  358  372  398  433  464  493  515  517   518  555  593  613  620  634  642  650  664  676  690  723  728  752  759   507
       5.5  5.8  6.7  6.7  7.0  7.0  7.0  7.1  7.1  7.3  7.4  7.6  7.7  7.8  7.8   7.8  8.0  8.2  8.3  8.3  8.4  8.4  8.4  8.5  8.6  8.7  9.0  9.0  9.2  9.2   7.8

Teita   80  199  211  232  354  384  420  448  456  460  472  488  494  507  521   535  548  565  569  571  573  614  621  654  659  669  731  743  771  800   512
       5.3  6.2  6.3  6.4  7.0  7.2  7.4  7.5  7.6  7.6  7.6  7.7  7.7  7.8  7.8   7.9  8.0  8.1  8.1  8.1  8.1  8.3  8.3  8.5  8.5  8.5  9.0  9.1  9.3  9.7   7.8

Zulu   184  223  240  302  314  350  370  403  422  429  452  468  479  497  506   524  532  551  579  583  591  597  661  697  706  729  741  750  761  814   515
       6.1  6.3  6.5  6.8  6.8  7.0  7.1  7.3  7.4  7.4  7.5  7.6  7.7  7.8  7.8   7.9  7.9  8.0  8.1  8.2  8.2  8.2  8.5  8.8  8.9  9.0  9.1  9.2  9.2   10   7.9

Tasman 141  213  279  298  352  355  364  375  399  419  455  480  487  489  501   528  534  552  558  566  580  619  640  730  736  791  817  822  828  830   525
       5.8  6.3  6.7  6.7  7.0  7.0  7.1  7.2  7.3  7.4  7.5  7.7  7.7  7.7  7.8   7.9  7.9  8.0  8.1  8.1  8.1  8.3  8.4  9.0  9.1  9.6   10   10   10   11   8.0

Bushma 169  305  308  309  328  331  361  385  400  409  412  454  481  491  539   541  568  587  610  632  662  666  686  710  714  715  720  779  794  809   529
       6.0  6.8  6.8  6.8  6.9  6.9  7.2  7.3  7.3  7.3  7.5  7.7  7.7  7.7  7.9   7.9  8.1  8.2  8.3  8.4  8.5  8.5  8.6  8.9  8.9  8.9  9.5  9.6  9.8  9.9   7.9

Hawaii 206  209  397  402  434  446  447  453  522  531  542  546  554  562  572   581  589  608  617  630  636  694  699  701  704  746  748  770  787  819   572
       6.2  6.3  7.3  7.3  7.4  7.5  7.5  7.5  7.9  7.9  7.9  8.0  8.0  8.1  8.1   8.1  8.2  8.3  8.3  8.4  8.4  8.8  8.8  8.8  8.8  9.1  9.1  9.3  9.5   10   8.2

Easter 180  243  391  406  459  462  474  490  514  520  530  533  537  549  585   592  601  625  627  653  688  695  705  735  745  749  754  785  793  812   581
       6.1  6.5  7.2  7.3  7.6  7.6  7.6  7.7  7.8  7.8  7.9  7.9  7.9  8.0  8.2   8.2  8.2  8.3  8.3  8.5  8.7  8.8  8.8  8.9  9.1  9.1  9.2  9.2  9.5  9.6  9.9  8.2

Morior 239  357  404  414  471  525  556  604  631  633  638  644  658  667  677   678  684  713  744  760  766  767  778  784  786  789  805  806  807  833   654
       6.5  7.1  7.3  7.3  7.6  7.9  8.0  8.3  8.4  8.4  8.4  8.4  8.5  8.5  8.6   8.6  8.6  8.9  9.1  9.2  9.3  9.3  9.4  9.5  9.5  9.5  9.8  9.8  9.8   11   8.7

Buriat 148  166  294  324  426  513  559  607  622  626  663  691  707  709  722   727  758  768  772  773  774  775  790  798  801  804  826  829  835  838   655
       5.8  5.9  6.7  6.9  7.4  7.8  8.1  8.3  8.3  8.3  8.5  8.7  8.9  8.9  9.0   9.0  9.2  9.3  9.3  9.4  9.4  9.4  9.6  9.6  9.7  9.7   10   10   11   11   8.8

Eskimo 266  424  443  538  547  605  612  629  635  645  649  655  660  683  698   703  712  734  738  762  763  777  781  788  792  795  803  813  823  837   677
       6.6  7.4  7.5  7.9  8.0  8.3  8.3  8.4  8.4  8.4  8.4  8.5  8.5  8.6  8.8   8.8  8.9  9.1  9.1  9.3  9.3  9.4  9.5  9.5  9.6  9.6  9.7   10   10   11   8.8

Dogon  299  310  377  381  484  575  584  641  648  657  692  718  719  721  726   739  742  751  757  765  769  776  782  797  808  810  824  825  827  831   678
       6.7  6.8  7.2  7.2  7.8  8.1  8.2  8.4  8.4  8.5  8.7  8.9  8.9  9.0  9.0   9.1  9.1  9.2  9.2  9.3  9.3  9.4  9.5  9.6  9.8  9.8   10   10   10   11   8.9

Austra 273  276  440  511  523  550  561  647  665  671  682  696  702  724  725   733  737  747  753  780  783  799  802  815  816  818  820  834  836  839   685
       6.7  6.7  7.5  7.8  7.9  8.0  8.1  8.4  8.5  8.6  8.6  8.8  8.8  9.0  9.0   9.0  9.1  9.1  9.2  9.5  9.5  9.6  9.7   10   10   10   10   11   11   12   9.0
```

more complex POPKIN display, the closeness of Easter is manifest, and of all 90 Polynesians in the control samples, only one individual falls outside the lowest rank moiety (those under 420 in rank), and within all three Polynesian populations almost all the nearer half (lowest 15) cases have individual distances at or below a value of 5.3. Clearly, this specimen lies in a sort of generalized Polynesian space, and least distant from the Easter centroid.

Compared to DISPOP, for #3029 Guam is displaced in general nearness by Moriori. This is hard to visualize geometrically[3], but while the nearest Guam skull has the rank overall of 22, this is followed by a long array of Guam specimens which are not very distant and in fact, at much of the middle range, less distant than Hawaii or Moriori. It is notable that Guam violates the tendency for the median of a population to reflect its mean distance in general. Guamanian individuals at the median range are distinctly closer to #3029 than in the case of Hawaii or Moriori. Viewed in one way (not very profitably) we might see lobes of these populations disposed in complicated ways in the hyperspace; viewed in another, we certainly see #3029 as approached closely in shape by given individual skulls, notably Polynesians, and in some more generalized way by Guamanian skulls, but with unspecified distinctions overall. This would be signaled by the lack of any really near specimens in spite of the closeness of the Guam series over the main range.

Also visible is the fact that the next most distant populations are all Asiatic or American ("Mongoloid," but excluding Buriats or Eskimos), all 11 such samples down to Ainu being included before Europeans or Africans make an appearance. The one interesting exception is made by the Melanesian Tolai.

Number 3022 (Table 9B-2) finds a less comfortable Polynesian home. Compared to DISPOP, Hawaii displaces Moriori as the stated nearest sample, but the P value in Table 3 is not high for

either, and in POPKIN only the first 34 skulls in rank, out of the whole 839, have an individual distance of 5.3 or less. Again, as in the previous case, the Ainu, though ranked behind Guam in mean distance, has a string of cases in the nearer 15 closer to the target than the Guam sample in this part of the range.

Again, blind judgment as to the location of #3022 would have to be "some kind of Polynesian," in a part of the universe where Guamanians and Ainus are also present. Again, all 11 of the specified Asian and American samples are broadly nearer to the target than is any other population.

Number 3027 disagrees with DISPOP (table 3) in diagnosis, where it was Moriori (P=.35) over Tolai (P=.16). Here assignment is positively Tolai, with 17 of 30 control Tolais having an individual distance of 5.3 or less. Australians and Tasmanians are in the near distance. Nevertheless, the nearer populations, led by Guam and Ainu and followed by Polynesians, are mainly Asiatic and American.

This case is the most egregious departure from expectation among these 20 known Maori TEST cases. In some others, there are also non-Polynesian attributions, in a few cases seemingly more positively than in the DISPOP resolution. Even so, there is a certain regularity. The nearer populations are consistently Asiatic and American (always excepting Buriats and Eskimos; see remarks as to these populations above, under "Centrality of Shape"). Tolais are within this perimeter in a number of cases, but not Australians or Tasmanians. Europeans are consistently distant, if indeed they appear among the 18 listed populations at all, and Africans appear there only rarely.

Certainly POPKIN is good evidence for general location of a target individual, and for exclusion of populations or regions from consideration. For satisfying identification of a target it is less successful but, again, when target and control population are not matched, it is not expected that such identification will be made consistently.

---

3. Positions cannot be visualized as if in a real space of three dimensions, since this is a hyperspace of 14 dimensions.

TABLE 9B-1
POPKIN results: Maori TEST #3029

```
Easter    2   6   7  12  18  20  24  29  33  41  44  47  51  54  79     99 120 122 126 138 142 149 158 179 186 232 248 257 277 300   107
        3.1 3.4 3.6 3.8 4.0 4.3 4.3 4.4 4.5 4.6 4.6 4.7 4.8 4.8 5.1    5.3 5.4 5.5 5.5 5.6 5.6 5.6 5.7 5.8 5.8 6.0 6.1 6.2 6.3 6.4   5.0

Hawaii    3   4   5   9  10  19  30  31  38  42  45 117 133 155 156    160 188 201 207 210 213 220 250 252 256 260 267 268 305 374   148
        3.2 3.2 3.4 3.6 3.8 4.1 4.4 4.5 4.6 4.6 4.7 5.4 5.5 5.6 5.7    5.7 5.8 5.9 5.9 5.9 6.0 6.0 6.1 6.1 6.1 6.2 6.2 6.2 6.4 6.7   5.3

Guam     22  32  36  37  55  57  63  65  69  74  76  86  90  95  98    124 146 161 174 190 196 208 211 215 303 310 361 367 426 621   166
        4.3 4.5 4.5 4.6 4.8 4.9 5.0 5.0 5.1 5.1 5.1 5.2 5.2 5.3 5.3    5.5 5.6 5.7 5.8 5.8 5.8 5.9 5.9 6.0 6.0 6.4 6.4 6.7 6.7 6.9 7.9   5.6

Morior    8  11  13  14  17  26  27  28  35  56  73  85  88 102 136    147 152 182 218 245 270 282 302 325 349 376 380 384 392 445   172
        3.6 3.8 3.9 4.0 4.0 4.4 4.4 4.4 4.5 4.9 5.1 5.2 5.2 5.3 5.6    5.6 5.6 5.8 6.0 6.1 6.2 6.3 6.4 6.5 6.6 6.7 6.8 6.8 6.8 7.0   5.4

Anyang   25  34  39  43  46  52  64  89 119 123 125 134 139 157 171    177 183 216 224 243 261 265 286 291 311 348 391 439 456 492   198
        4.4 4.5 4.6 4.6 4.7 4.8 5.0 5.2 5.4 5.5 5.5 5.5 5.6 5.7 5.7    5.8 5.8 6.0 6.0 6.1 6.2 6.2 6.3 6.3 6.4 6.6 6.8 7.0 7.1 7.3   5.7

Arikar    1  59  61  67  70  71  72  96 100 103 114 180 194 195 199    203 221 253 278 336 341 351 354 395 414 422 461 466 481 657   241
        3.0 4.9 4.9 5.0 5.1 5.1 5.1 5.3 5.3 5.3 5.4 5.8 5.9 5.9 5.9    5.9 6.0 6.1 6.3 6.5 6.6 6.6 6.6 6.8 6.9 6.9 7.1 7.1 7.2 8.2   6.0

Peru     53  58  60  75  80  82 107 143 144 151 154 169 184 200 204    247 262 269 280 295 312 314 337 416 417 442 467 469 512 654   252
        4.8 4.9 4.9 5.1 5.1 5.1 5.4 5.6 5.6 5.6 5.6 5.7 5.8 5.9 5.9    6.1 6.2 6.2 6.3 6.4 6.4 6.4 6.4 6.5 6.9 6.9 7.0 7.1 7.1 7.4 8.2   6.1

SJapan   23  40  62  66  77 109 110 116 118 128 130 141 163 167 233    309 330 338 339 365 394 401 420 435 465 474 518 566 588 709   278
        4.3 4.6 4.9 5.0 5.1 5.4 5.4 5.4 5.4 5.5 5.5 5.6 5.7 5.7 6.0    6.4 6.5 6.6 6.6 6.7 6.8 6.9 6.9 7.0 7.1 7.2 7.4 7.6 7.7 8.6   6.2

Hainan   83  93 121 148 153 170 178 185 225 241 283 284 285 290 306    308 317 321 322 333 335 356 377 386 396 408 433 441 487 705   299
        5.1 5.2 5.5 5.6 5.6 5.7 5.8 5.8 6.0 6.1 6.3 6.3 6.3 6.3 6.4    6.4 6.5 6.5 6.5 6.5 6.5 6.6 6.8 6.8 6.8 6.9 7.0 7.0 7.2 8.5   6.4

NJapan   16  21  49  68  81 131 132 140 162 192 202 231 246 259 292    299 319 334 344 345 402 415 418 423 437 499 626 663 666 685   303
        4.0 4.3 4.7 5.0 5.1 5.5 5.5 5.6 5.7 5.9 5.9 6.0 6.1 6.2 6.3    6.4 6.5 6.5 6.6 6.6 6.6 6.9 6.9 6.9 6.9 7.0 7.3 8.0 8.3 8.3 8.4   6.3

Philip   50  78  97 115 137 150 164 197 198 234 244 254 255 288 315    327 346 347 353 355 438 464 484 485 490 547 553 604 625 648   328
        4.8 5.1 5.3 5.4 5.6 5.6 5.7 5.9 5.9 6.0 6.1 6.1 6.1 6.3 6.5    6.5 6.6 6.6 6.6 6.6 6.7 7.1 7.2 7.2 7.2 7.5 7.5 7.8 8.0 8.1   6.5

Atayal  113 173 175 217 238 273 279 281 297 304 320 350 358 363 364    368 375 412 421 429 431 432 443 447 489 502 514 529 638   .    363
        5.4 5.8 5.8 6.0 6.1 6.2 6.3 6.3 6.4 6.4 6.5 6.6 6.6 6.7 6.7    6.7 6.7 6.9 6.9 7.0 7.0 7.0 7.0 7.0 7.0 7.2 7.3 7.4 7.5 8.0   .    6.7

Tolai    15  91 105 129 165 166 251 258 276 293 298 372 407 434 444    449 457 475 477 480 522 530 531 542 551 595 623 627 665 671   397
        4.0 5.2 5.3 5.5 5.7 5.7 6.1 6.2 6.3 6.3 6.4 6.7 6.9 7.0 7.0    7.1 7.1 7.2 7.2 7.2 7.4 7.5 7.5 7.5 7.5 7.8 8.0 8.0 8.3 8.3   6.8

S.Cruz   87 106 111 145 228 237 242 264 266 296 316 329 331 366 370    378 405 458 491 493 495 520 534 535 600 614 619 649 711 742   398
        5.2 5.4 5.4 5.6 6.0 6.1 6.1 6.2 6.2 6.4 6.5 6.5 6.5 6.7 6.7    6.8 6.9 7.1 7.2 7.3 7.3 7.4 7.5 7.5 7.8 7.9 7.9 8.1 8.6 9.0   6.9

Ainu     84 104 108 127 189 222 230 236 271 272 369 387 390 399 468    479 486 508 560 573 579 585 586 590 591 633 644 651 697 727   425
        5.2 5.3 5.4 5.5 5.8 6.0 6.0 6.1 6.2 6.2 6.7 6.8 6.8 6.8 7.1    7.2 7.2 7.4 7.6 7.6 7.7 7.7 7.7 7.7 7.8 8.0 8.1 8.1 8.5 8.8   7.0

Zalava  181 205 206 214 223 239 240 249 318 323 326 340 389 400 403    404 450 497 500 513 532 563 568 583 598 612 710 712 714 724   431
        5.8 5.9 5.9 6.0 6.1 6.1 6.1 6.1 6.5 6.5 6.5 6.6 6.8 6.9 6.9    6.9 7.1 7.3 7.3 7.4 7.5 7.6 7.6 7.7 7.8 7.9 8.6 8.6 8.6 8.7   7.0

Norse    94 101 112 172 187 294 332 359 373 398 411 448 459 473 476    505 515 537 555 557 559 606 646 667 669 676 678 720 735 770   469
        5.2 5.3 5.4 5.7 5.8 6.3 6.5 6.7 6.7 6.8 6.9 7.1 7.1 7.2 7.2    7.3 7.4 7.5 7.6 7.6 7.6 7.9 8.1 8.3 8.3 8.3 8.3 8.7 8.9 9.4   7.2

Teita    92 193 212 263 289 352 360 383 388 397 419 424 436 451 460    472 501 545 549 575 592 601 605 643 680 683 686 700 736 752   481
        5.2 5.9 6.0 6.2 6.3 6.6 6.7 6.8 6.8 6.8 6.9 6.9 7.0 7.1 7.1    7.1 7.3 7.5 7.5 7.7 7.7 7.8 7.8 8.1 8.4 8.4 8.4 8.5 9.0 9.1   7.3

Eskimo  168 176 209 235 274 301 307 328 343 425 430 453 517 521 524    525 540 543 562 567 576 593 615 650 681 694 695 739 757 801   492
        5.7 5.8 5.9 6.1 6.2 6.4 6.4 6.5 6.6 6.9 7.0 7.1 7.4 7.4 7.5    7.5 7.5 7.5 7.6 7.6 7.7 7.7 7.9 8.1 8.4 8.4 8.5 9.0 9.2 9.8   7.4

Zulu    219 229 362 381 427 440 452 506 509 510 519 526 536 539 544    548 554 599 631 647 656 658 661 674 677 682 687 699 722 832   554
        6.0 6.0 6.7 6.8 6.9 7.0 7.1 7.3 7.4 7.4 7.4 7.5 7.5 7.5 7.5    7.5 7.5 7.8 8.0 8.1 8.2 8.2 8.2 8.3 8.3 8.4 8.4 8.5 8.7  11    7.7

Egypt   191 275 313 371 379 446 454 455 462 478 482 494 546 556 558    569 602 608 620 641 642 689 704 728 730 737 738 759 760 766   558
        5.9 6.2 6.4 6.7 6.8 7.0 7.1 7.1 7.2 7.2 7.3 7.5 7.6 7.6 7.6    7.6 7.8 7.9 7.9 8.1 8.1 8.4 8.5 8.8 8.8 9.0 9.0 9.2 9.2 9.3   7.7

Andama  135 227 342 409 428 463 488 507 523 533 538 550 552 571 574    577 580 589 597 611 616 630 664 673 675 684 703 769 776 802   560
        5.6 6.0 6.6 6.9 7.0 7.1 7.2 7.4 7.5 7.5 7.5 7.5 7.5 7.6 7.7    7.7 7.7 7.7 7.8 7.9 7.9 8.0 8.3 8.3 8.3 8.4 8.5 9.3 9.4 9.8   7.7

Berg    226 357 382 385 410 413 470 471 483 498 503 516 565 618 628    636 659 660 668 691 696 729 731 753 771 772 783 786 804 824   596
        6.0 6.6 6.8 6.8 6.9 6.9 7.1 7.1 7.2 7.3 7.3 7.4 7.6 7.9 8.0    8.0 8.2 8.2 8.3 8.4 8.5 8.8 8.9 9.1 9.4 9.4 9.5 9.6 9.8  10    8.0

Tasman   48 159 287 406 496 504 570 572 582 613 632 639 653 662 672    716 717 723 725 745 746 751 763 767 774 789 790 792 819 835   632
        4.7 5.7 6.3 6.9 7.3 7.3 7.6 7.6 7.7 7.9 8.0 8.1 8.2 8.2 8.3    8.6 8.7 8.8 8.9 9.1 9.1 9.1 9.2 9.3 9.4 9.6 9.6 9.7  10  12    8.3

Austra  324 393 527 581 603 609 610 617 622 624 635 640 655 679 690    701 706 713 719 732 733 755 756 761 785 791 807 811 817 831   674
        6.5 6.8 7.5 7.7 7.8 7.9 7.9 7.9 7.9 8.0 8.0 8.1 8.2 8.3 8.4    8.5 8.6 8.6 8.7 8.9 8.9 9.1 9.2 9.2 9.6 9.6 9.9  10  10  11    8.6

Buriat  511 541 594 596 629 634 688 702 708 718 740 747 750 765 778    781 782 784 787 795 799 810 812 820 826 828 829 833 836 838   742
        7.4 7.5 7.7 7.8 8.0 8.0 8.4 8.5 8.6 8.7 9.0 9.1 9.1 9.3 9.4    9.4 9.4 9.5 9.6 9.7 9.8  10  10  10  11  11  11  11  12  12    9.4

Bushma  561 564 584 652 692 693 707 715 721 726 743 744 748 749 754    764 773 777 794 797 798 803 805 813 814 815 816 821 834 839   747
        7.6 7.6 7.7 8.2 8.4 8.4 8.6 8.6 8.7 8.8 9.0 9.1 9.1`9.1 9.1    9.3 9.4 9.4 9.7 9.7 9.8 9.8 9.9  10  10  10  10  10  11  12    9.3

Dogon   528 578 587 607 637 645 670 698 734 741 758 762 768 775 779    780 788 793 796 800 806 808 809 818 822 823 825 827 830 837   748
        7.5 7.7 7.7 7.9 8.0 8.1 8.3 8.5 8.9 9.0 9.2 9.2 9.3 9.4 9.4    9.4 9.6 9.7 9.7 9.8 9.9 9.9 9.9  10  10  10  11  11  11  12    9.4
```

TABLE 9B-2

POPKIN results: Maori TEST #3022

Hawaii   5   6   9 20 21 26 29 37 39 44 47 48 60 61 66   74 90 99 105 108 118 138 146 177 215 260 273 389 410 452 119
4.5 4.6 4.9 5.1 5.1 5.2 5.3 5.4 5.4 5.6 5.6 5.6 5.8 5.8 5.9 5.9 6.0 6.1 6.2 6.2 6.3 6.4 6.5 6.7 6.9 7.2 7.2 7.8 7.9 8.1 6.0

Morior   1   2   3 10 22 30 32 34 35 49 51 53 59 63 79   96 114 131 133 165 166 202 207 212 218 346 421 441 448 458 143
3.9 4.0 4.1 4.9 5.2 5.3 5.3 5.3 5.4 5.6 5.7 5.7 5.8 5.8 6.0 6.1 6.2 6.4 6.4 6.6 6.6 6.8 6.9 6.9 6.9 7.6 7.9 8.0 8.1 8.2 6.1

Guam   7 15 18 28 36 73 88 102 106 116 120 126 135 144 145 149 153 158 167 185 203 209 216 232 233 251 282 288 296 300 153
4.7 5.0 5.1 5.3 5.4 5.9 6.0 6.2 6.2 6.3 6.3 6.3 6.4 6.5 6.5 6.5 6.5 6.5 6.6 6.7 6.8 6.9 6.9 7.0 7.0 7.1 7.3 7.3 7.4 7.4 6.4

Ainu   12 14 23 25 33 46 62 72 76 77 85 89 112 115 125 141 147 174 191 199 213 221 241 262 321 330 343 399 438 484 169
4.9 5.0 5.2 5.2 5.3 5.6 5.8 5.9 5.9 5.9 6.0 6.0 6.2 6.3 6.3 6.4 6.5 6.7 6.7 6.8 6.9 6.9 7.1 7.2 7.5 7.5 7.6 7.8 8.0 8.3 6.5

Arikar   8 11 27 31 83 84 87 98 103 117 140 160 161 168 179 192 210 226 245 254 263 286 292 311 314 327 337 366 430 437 198
4.9 4.9 5.2 5.3 6.0 6.0 6.0 6.1 6.2 6.3 6.4 6.6 6.6 6.6 6.7 6.7 6.9 7.0 7.1 7.1 7.2 7.3 7.3 7.4 7.4 7.5 7.6 7.8 8.0 8.0 6.7

Hainan 19 86 97 100 101 121 132 134 150 176 180 195 211 225 229 237 238 239 243 247 331 338 352 364 367 398 432 442 568 570 251
5.1 6.0 6.1 6.1 6.2 6.3 6.4 6.4 6.5 6.7 6.7 6.8 6.9 7.0 7.0 7.0 7.0 7.0 7.1 7.1 7.5 7.6 7.6 7.7 7.7 7.7 7.8 8.0 8.8 8.9 7.0

Easter 41 43 50 82 113 157 171 175 182 205 208 214 224 231 248 253 264 270 275 298 333 341 351 413 431 457 465 495 507 574 269
5.5 5.5 5.6 6.0 6.2 6.5 6.6 6.7 6.7 6.8 6.9 6.9 7.0 7.0 7.1 7.1 7.2 7.2 7.2 7.4 7.6 7.6 7.6 7.9 8.0 8.1 8.2 8.4 8.5 8.9 7.1

NJapan 16 38 40 56 71 78 91 123 148 151 154 162 169 230 244 256 259 303 304 309 320 329 412 454 482 499 530 581 627 636 269
5.0 5.4 5.5 5.7 5.9 6.0 6.0 6.3 6.5 6.5 6.5 6.6 6.6 7.0 7.1 7.2 7.2 7.4 7.4 7.4 7.5 7.5 7.9 8.1 8.3 8.4 8.6 8.9 9.3 9.3 7.1

Anyang 17 67 69 75 81 137 163 178 181 183 186 242 267 277 284 302 305 317 319 328 336 374 384 414 423 433 478 509 526 584 281
5.0 5.9 5.9 5.9 6.0 6.4 6.6 6.7 6.7 6.7 6.7 7.1 7.2 7.3 7.3 7.4 7.4 7.5 7.5 7.5 7.6 7.7 7.8 7.9 7.9 8.0 8.3 8.5 8.6 8.9 7.2

Philip 13 52 122 129 136 142 152 156 159 189 198 219 255 258 271 294 295 323 332 347 358 372 377 381 427 446 480 489 546 572 283
4.9 5.7 6.3 6.4 6.4 6.5 6.5 6.5 6.6 6.7 6.8 6.9 7.1 7.2 7.2 7.3 7.4 7.5 7.5 7.6 7.7 7.7 7.7 7.8 8.0 8.1 8.3 8.4 8.7 8.9 7.2

Peru   4 42 45 64 80 127 130 143 170 200 222 227 234 266 268 307 326 362 363 385 416 426 444 468 469 488 502 525 608 610 297
4.5 5.5 5.6 5.8 6.0 6.3 6.4 6.5 6.6 6.8 7.0 7.0 7.0 7.2 7.2 7.4 7.5 7.7 7.7 7.8 7.9 8.0 8.0 8.2 8.2 8.4 8.5 8.6 9.1 9.1 7.2

S.Cruz 55 65 109 124 128 155 187 193 194 197 201 217 228 236 240 280 285 316 318 357 391 397 402 411 422 455 519 536 646 664 298
5.7 5.8 6.2 6.3 6.4 6.5 6.7 6.7 6.8 6.8 6.8 6.9 7.0 7.0 7.1 7.3 7.3 7.5 7.5 7.7 7.8 7.8 7.8 7.9 7.9 8.1 8.5 8.6 9.3 9.5 7.3

SJapan 58 92 94 110 111 119 173 188 220 278 291 297 301 306 324 335 344 350 376 392 393 401 403 490 513 520 564 575 590 602 327
5.8 6.1 6.1 6.2 6.2 6.3 6.6 6.7 6.9 7.3 7.3 7.4 7.4 7.4 7.5 7.6 7.6 7.6 7.7 7.8 7.8 7.8 7.8 8.4 8.5 8.5 8.8 8.9 9.0 9.1 7.5

Atayal 56 107 166 190 204 246 249 257 281 290 299 312 313 315 334 339 371 373 380 388 415 440 451 463 464 486 517 547 579   . 339
5.7 6.2 6.6 6.7 6.8 7.1 7.1 7.2 7.3 7.3 7.4 7.4 7.4 7.4 7.6 7.6 7.7 7.7 7.8 7.8 7.9 8.0 8.1 8.2 8.2 8.4 8.5 8.7 8.9   . 7.8

Andama 70 104 172 283 342 349 368 382 386 417 418 420 429 456 460 461 466 472 485 487 524 531 533 544 576 577 598 704 746 778 451
5.9 6.2 6.6 7.3 7.6 7.6 7.7 7.8 7.8 7.9 7.9 7.9 8.0 8.1 8.2 8.2 8.2 8.3 8.3 8.4 8.6 8.6 8.7 8.9 8.9 9.0 9.8 10 11 8.2

Zalava 93 139 184 252 287 340 369 383 395 400 434 467 483 494 496 501 505 506 514 550 561 566 567 617 632 675 680 682 693 697 472
6.1 6.4 6.7 7.1 7.3 7.6 7.7 7.8 7.8 7.8 8.0 8.2 8.3 8.4 8.4 8.5 8.5 8.5 8.5 8.7 8.8 8.8 8.8 8.9 9.2 9.3 9.6 9.6 9.6 9.7 9.8 8.3

Tolai   24 57 235 265 345 361 365 378 406 439 445 459 492 511 512 518 522 523 538 541 549 588 649 657 672 699 719 733 774 775 492
5.2 5.7 7.0 7.2 7.6 7.7 7.7 7.7 7.7 7.9 8.0 8.0 8.2 8.4 8.5 8.5 8.5 8.6 8.7 8.7 8.9 9.0 9.4 9.4 9.5 9.8 10 10 11 11 8.4

Norse 95 196 206 276 279 289 322 348 354 370 435 497 500 527 543 545 552 555 569 653 659 670 673 692 694 705 715 742 745 754 505
6.1 6.8 6.8 7.2 7.3 7.3 7.5 7.6 7.6 7.7 8.0 8.4 8.4 8.6 8.7 8.7 8.8 8.8 8.9 9.4 9.4 9.5 9.6 9.7 9.7 9.8 9.9 10 10 10 8.6

Berg 261 269 308 356 408 419 424 425 436 443 479 504 542 578 589 594 612 613 616 624 650 663 678 688 696 702 716 736 744 757 551
7.2 7.2 7.4 7.7 7.9 7.9 7.9 7.9 8.0 8.0 8.3 8.5 8.7 8.9 9.0 9.0 9.2 9.2 9.2 9.2 9.4 9.4 9.6 9.7 9.7 9.8 9.9 10 10 10 8.8

Egypt 272 274 310 375 387 390 394 407 462 477 503 532 559 592 593 600 606 614 621 629 630 641 647 661 691 709 740 762 769 776 554
7.2 7.2 7.4 7.7 7.8 7.8 7.8 7.9 8.2 8.3 8.5 8.6 8.8 9.0 9.0 9.0 9.1 9.2 9.2 9.3 9.3 9.4 9.4 9.7 9.8 10 10 10 11 11 8.9

Eskimo 223 250 355 360 379 447 450 474 498 510 535 540 551 553 573 582 585 597 611 638 644 658 660 676 739 755 765 785 786 804 566
7.0 7.1 7.7 7.7 7.8 8.1 8.1 8.3 8.4 8.5 8.6 8.7 8.8 8.8 8.9 8.9 8.9 9.0 9.2 9.3 9.3 9.4 9.4 9.6 10 10 10 11 11 11 9.0

Tasman 68 293 325 353 404 405 428 493 508 571 583 601 603 620 625 634 637 652 679 684 701 711 729 737 741 748 758 809 813 835 592
5.9 7.3 7.5 7.6 7.8 7.8 8.0 8.4 8.5 8.9 8.9 9.0 9.1 9.2 9.3 9.3 9.3 9.4 9.6 9.7 9.8 9.9 10 10 10 10 10 11 11 13 9.2

Zulu 359 396 449 475 521 534 539 554 563 591 596 599 607 618 631 633 635 640 648 665 669 685 712 713 714 743 747 759 781 815 620
7.7 7.8 8.1 8.3 8.5 8.6 8.7 8.8 8.8 9.0 9.0 9.0 9.1 9.2 9.3 9.3 9.3 9.3 9.4 9.5 9.5 9.7 9.9 9.9 9.9 10 10 10 11 11 9.3

Austra 409 453 470 473 516 528 529 548 557 560 562 586 605 639 643 654 668 681 690 703 710 727 732 768 770 777 787 795 825 828 640
7.9 8.1 8.2 8.3 8.5 8.6 8.6 8.7 8.8 8.8 8.9 9.0 9.1 9.3 9.3 9.4 9.5 9.6 9.7 9.8 9.9 10 10 11 11 11 11 11 12 12 9.5

Buriat 471 481 558 609 615 622 623 626 655 656 687 695 698 722 726 734 749 750 756 761 763 767 788 790 802 810 817 818 824 837 707
8.2 8.3 8.8 9.1 9.2 9.2 9.2 9.3 9.4 9.4 9.7 9.7 9.8 10 10 10 10 10 10 10 11 11 11 11 11 11 11 12 12 13 10.1

Teita 491 515 565 587 595 628 642 667 671 674 677 700 718 720 723 724 725 728 752 773 782 783 784 793 801 803 806 807 812 819 709
8.4 8.5 8.8 9.0 9.0 9.3 9.3 9.5 9.5 9.6 9.6 9.8 10 10 10 10 10 10 10 11 11 11 11 11 11 11 11 11 11 12 10.1

Dogon 476 537 556 580 604 619 645 662 666 686 689 706 708 717 731 735 753 760 764 772 779 789 791 794 796 797 798 808 816 830 712
8.3 8.6 8.8 8.9 9.1 9.2 9.3 9.4 9.5 9.7 9.7 9.8 9.8 9.9 10 10 10 10 10 11 11 11 11 11 11 11 11 11 12 12 10.1

Bushma 651 683 707 721 730 738 751 766 771 780 792 799 800 805 811 814 820 821 822 823 826 827 829 831 832 833 834 836 838 839 791
9.4 9.7 9.8 10 10 10 10 10 11 11 11 11 11 11 11 11 12 12 12 12 12 12 12 12 12 12 12 13 13 13 11.3

TABLE 9B-3

POPKIN results: Maori TEST #3027

```
Tolai     1    2    5    6   14   20   26   27   28   34   35   38   39   48   74     87   95  116  132  149  150  173  188  199  214  290  364  400  419  454   128
        2.7  3.6  3.8  3.9  4.2  4.4  4.6  4.6  4.6  4.8  4.8  4.8  4.9  5.0  5.2    5.3  5.3  5.5  5.6  5.6  5.6  5.7  5.8  5.8  5.9  6.2  6.4  6.6  6.7  6.8   5.2

Guam     13   31   40   41   58   61   72   90   91   97  105  128  143  197  212    223  235  238  252  253  283  311  336  340  343  401  474  483  553  617   224
        4.2  4.7  4.9  4.9  5.1  5.1  5.2  5.3  5.3  5.4  5.4  5.5  5.6  5.8  5.9    5.9  6.0  6.0  6.0  6.0  6.2  6.2  6.3  6.3  6.3  6.6  6.8  6.9  7.1  7.4   5.8

Ainu      9   15   22   57   68   69   79   98  103  118  124  137  139  204  206    237  271  277  280  284  319  335  341  348  354  396  417  475  611  638   234
        4.1  4.3  4.5  5.1  5.2  5.2  5.2  5.4  5.4  5.5  5.5  5.6  5.6  5.9  5.9    6.0  6.1  6.1  6.1  6.2  6.3  6.3  6.3  6.4  6.4  6.5  6.7  6.8  7.4  7.5   5.8

Morior   11   23   29   37   52   86  114  115  117  140  146  156  157  191  193    195  210  234  266  268  358  377  383  456  458  511  512  583  608  641   254
        4.1  4.5  4.7  4.8  5.0  5.3  5.4  5.5  5.5  5.6  5.6  5.6  5.6  5.8  5.8    5.8  5.9  6.0  6.1  6.1  6.4  6.5  6.5  6.8  6.8  7.0  7.0  7.2  7.4  7.5   5.9

Arikar   24   43   64   80   88   94   99  100  107  142  166  169  174  187  251    297  306  320  323  380  381  390  455  493  498  499  504  517  582  693   281
        4.5  4.9  5.2  5.3  5.3  5.3  5.4  5.4  5.4  5.6  5.7  5.7  5.7  5.8  6.0    6.2  6.2  6.3  6.3  6.5  6.5  6.5  6.8  6.9  6.9  6.9  7.0  7.0  7.2  7.9   6.1

Hawaii    8   19   36  122  134  155  158  179  186  192  194  196  200  211  213    289  305  314  329  332  369  371  375  389  397  398  428  599  637  728   282
        4.1  4.4  4.8  5.5  5.6  5.6  5.7  5.7  5.8  5.8  5.8  5.8  5.8  5.9  5.9    6.2  6.2  6.2  6.3  6.3  6.5  6.5  6.5  6.5  6.6  6.6  6.7  7.3  7.5  8.1   6.1

S.Cruz   25   44   49   54   59   66   76   81  111  121  123  129  172  175  222    256  260  262  296  303  327  376  461  485  558  612  655  683  711  714   282
        4.5  5.0  5.0  5.1  5.1  5.2  5.2  5.3  5.4  5.5  5.5  5.5  5.7  5.7  5.9    6.1  6.1  6.1  6.2  6.2  6.3  6.5  6.8  6.9  7.1  7.4  7.6  7.8  8.0  8.0   6.1

Easter    4   16   21   46   65   77   83  130  198  201  209  227  228  245  254    301  308  353  407  410  436  441  471  491  492  521  541  542  550  645   297
        3.8  4.3  4.5  5.0  5.2  5.2  5.3  5.6  5.8  5.9  5.9  5.9  5.9  6.0  6.1    6.2  6.2  6.4  6.6  6.6  6.7  6.7  6.8  6.9  6.9  7.0  7.1  7.1  7.1  7.6   6.1

NJapan   10   45   55   62   63   70   96  119  145  167  176  185  221  257  261    275  278  281  317  328  384  392  415  432  505  609  659  730  744  746   301
        4.1  5.0  5.1  5.1  5.2  5.2  5.3  5.5  5.6  5.7  5.7  5.8  5.9  6.1  6.1    6.1  6.1  6.1  6.3  6.3  6.5  6.5  6.6  6.7  7.0  7.4  7.6  8.1  8.3  8.3   6.2

Austra   30   33   42   78   82   92  110  125  136  182  183  207  267  270  298    315  322  326  342  347  399  409  416  443  459  544  557  628  661  771   302
        4.7  4.7  4.9  5.2  5.3  5.3  5.4  5.5  5.6  5.8  5.8  5.9  6.1  6.1  6.2    6.2  6.3  6.3  6.3  6.4  6.6  6.6  6.6  6.8  6.8  7.1  7.1  7.5  7.7  8.6   6.2

Peru     50   73   84  101  102  106  147  148  164  165  230  231  242  248  265    292  325  344  365  368  423  429  438  448  462  476  508  604  660  664   305
        5.0  5.2  5.3  5.4  5.4  5.4  5.6  5.6  5.7  5.7  5.9  6.0  6.0  6.0  6.1    6.2  6.3  6.3  6.4  6.4  6.7  6.7  6.7  6.8  6.8  6.8  7.0  7.3  7.6  7.7   6.2

Zalava   60   71   85   93  113  126  159  162  205  219  236  240  263  304  362    393  404  411  442  447  466  481  497  547  560  567  598  650  718  731   354
        5.1  5.2  5.3  5.3  5.4  5.5  5.7  5.7  5.9  5.9  6.0  6.0  6.1  6.2  6.4    6.5  6.6  6.6  6.8  6.8  6.8  6.9  6.9  7.1  7.2  7.2  7.3  7.6  8.0  8.1   6.4

Tasman    3    7   12   51   89  135  202  217  229  279  293  300  313  318  351    372  440  446  486  500  528  533  545  622  634  670  681  686  701  799   381
        3.7  4.0  4.2  5.0  5.3  5.6  5.9  5.9  5.9  6.1  6.2  6.2  6.2  6.3  6.4    6.5  6.7  6.8  6.9  7.0  7.0  7.1  7.1  7.4  7.5  7.7  7.8  7.8  7.9  9.2   6.4

Philip  104  108  131  133  141  144  168  224  243  250  258  273  330  333  370    378  422  445  495  515  523  580  619  635  646  653  665  679  689  713   396
        5.4  5.4  5.6  5.6  5.6  5.6  5.7  5.9  6.0  6.1  6.1  6.3  6.3  6.5  6.5    6.5  6.7  6.8  6.9  7.0  7.2  7.4  7.5  7.6  7.6  7.7  7.7  7.8  7.8  8.0   6.6

Zulu     32  154  163  180  189  244  249  264  285  339  350  427  435  449  453    479  480  514  518  540  556  566  574  586  591  593  658  668  737  795   436
        4.7  5.6  5.7  5.7  5.8  6.0  6.0  6.1  6.2  6.3  6.4  6.7  6.7  6.8  6.8    6.9  6.9  7.0  7.0  7.1  7.1  7.2  7.2  7.3  7.3  7.3  7.6  7.7  8.2  9.0   6.7

Atayal  181  225  233  255  272  302  321  357  363  374  379  403  425  433  450    464  467  469  488  507  527  563  568  569  578  605  629  696  706    .   442
        5.7  5.9  6.0  6.1  6.1  6.2  6.3  6.4  6.4  6.5  6.5  6.6  6.7  6.7  6.8    6.8  6.8  6.8  6.9  7.0  7.0  7.2  7.2  7.2  7.2  7.3  7.5  7.9  7.9    .   6.7

Norse    47   56  109  153  190  216  274  331  345  373  382  385  408  414  451    482  510  519  526  535  571  579  614  647  700  725  741  749  763  768   452
        5.0  5.1  5.4  5.6  5.8  5.9  6.1  6.3  6.3  6.5  6.5  6.5  6.6  6.6  6.8    6.9  7.0  7.0  7.0  7.1  7.2  7.2  7.4  7.6  7.9  8.1  8.2  8.4  8.5  8.6   6.8

Hainan   53  120  138  208  220  286  294  346  352  356  361  418  437  484  489    529  539  546  552  554  561  565  572  592  621  663  680  710  751  775   459
        5.1  5.5  5.6  5.9  5.9  6.2  6.2  6.4  6.4  6.4  6.4  6.7  6.7  6.9  6.9    7.0  7.1  7.1  7.1  7.1  7.2  7.2  7.2  7.3  7.4  7.7  7.8  8.0  8.4  8.7   6.8

SJapan   67  127  170  177  203  269  299  316  334  337  367  395  402  452  468    490  509  516  534  551  581  595  610  662  692  707  723  764  781  814   464
        5.2  5.5  5.7  5.7  5.9  6.1  6.2  6.3  6.3  6.3  6.4  6.5  6.6  6.8  6.8    6.9  7.0  7.0  7.1  7.1  7.2  7.3  7.4  7.7  7.9  8.0  8.5  8.8  8.9  9.4   6.9

Eskimo  151  184  239  246  259  282  295  309  349  360  387  405  406  413  470    472  524  548  600  616  636  666  676  690  703  705  748  758  766  787   488
        5.6  5.8  6.0  6.0  6.1  6.1  6.2  6.2  6.4  6.4  6.5  6.6  6.6  6.6  6.8    6.8  7.0  7.1  7.3  7.4  7.5  7.7  7.8  7.8  7.9  7.9  8.4  8.5  8.6  8.9   7.0

Anyang   17   18  161  178  218  226  324  421  426  487  496  501  503  525  537    538  549  584  585  613  644  652  656  657  667  691  712  717  735  743   493
        4.4  4.4  5.7  5.7  5.9  5.9  6.3  6.7  6.7  6.9  6.9  7.0  7.0  7.0  7.1    7.1  7.1  7.2  7.2  7.4  7.6  7.6  7.6  7.6  7.7  7.8  8.0  8.0  8.1  8.3   6.9

Teita   112  152  232  247  307  310  312  359  420  424  460  520  522  532  575    590  594  607  620  639  651  677  682  715  721  752  777  782  784  786   529
        5.4  5.6  6.0  6.0  6.2  6.2  6.2  6.4  6.7  6.7  6.8  7.0  7.0  7.1  7.2    7.3  7.3  7.4  7.4  7.5  7.6  7.8  7.8  8.0  8.0  8.4  8.7  8.8  8.9  8.9   7.2

Berg     75  171  241  276  287  288  355  366  388  431  463  494  513  562  570    587  588  615  640  648  678  687  724  734  736  750  756  761  790  817   533
        5.2  5.7  6.0  6.1  6.2  6.2  6.4  6.4  6.5  6.7  6.8  6.9  7.0  7.2  7.2    7.3  7.3  7.4  7.5  7.6  7.8  7.8  8.0  8.1  8.1  8.4  8.4  8.5  8.9  9.4   7.2

Egypt   160  215  291  394  434  444  457  465  473  478  531  536  555  573  577    602  618  624  642  649  675  697  709  722  740  754  759  765  770  789   570
        5.7  5.9  6.2  6.5  6.7  6.8  6.8  6.8  6.8  6.9  7.1  7.1  7.1  7.2  7.2    7.3  7.4  7.4  7.6  7.6  7.8  7.9  8.0  8.0  8.2  8.4  8.5  8.5  8.6  8.9   7.4

Andama  338  391  412  477  559  576  589  596  597  603  606  623  630  643  669    674  716  719  729  732  733  742  747  757  767  769  773  778  794  815   652
        6.3  6.5  6.6  6.9  7.2  7.2  7.3  7.3  7.3  7.3  7.3  7.4  7.5  7.6  7.7    7.8  8.0  8.0  8.1  8.1  8.1  8.2  8.4  8.5  8.6  8.6  8.6  8.7  9.0  9.4   7.8

Dogon   430  439  502  506  530  543  632  673  684  685  695  698  702  704  720    727  753  760  772  783  788  793  805  806  810  812  821  828  834  837   702
        6.7  6.7  7.0  7.0  7.0  7.1  7.5  7.7  7.8  7.8  7.9  7.9  7.9  7.9  8.0    8.1  8.4  8.5  8.6  8.8  8.9  9.0  9.3  9.3  9.4  9.4  9.6   10   10   11   8.3

Buriat  386  564  631  633  671  688  726  738  739  755  762  776  779  785  791    792  801  803  804  808  816  823  825  826  827  831  832  833  836  839   757
        6.5  7.2  7.5  7.5  7.7  7.8  8.1  8.2  8.2  8.4  8.5  8.7  8.8  8.9  8.9    8.9  9.2  9.2  9.2  9.3  9.4  9.7  9.9  9.9  9.9   10   10   10   11   12   9.0

Bushma  601  625  626  627  654  672  694  699  708  745  774  780  796  797  798    800  802  807  809  811  813  818  819  820  822  824  829  830  835  838   762
        7.3  7.4  7.4  7.6  7.7  7.7  7.9  7.9  8.0  8.3  8.7  8.7  9.1  9.1  9.2    9.2  9.2  9.3  9.3  9.4  9.4  9.5  9.5  9.5  9.6  9.8   10   10   11   11   8.9
```

Late Prehistoric

First, I will mention a few particular cases with interesting results.

The two south Siberian Iron Age Kurgan (Scythian) skulls are most positively European in character, a character which is not that of the present inhabitants of the same region.

The Proto-Neolithic Natufian, *#1775*, is close to Zalavár, but not to Norse, whatever that may mean. In fact, it is approached by a number of controls from Asia, Africa, and the Southwest Pacific, seeming to lie in a central region, with 102 controls not further than a distance of 5.3. Nonetheless, Zalavár is clearly closest.

Fish Hoek of South Africa is latest Pleistocene. As with DISPOP, POPKIN makes a definite statement. Fish Hoek is decidedly Bushman, not any other kind of African: only 4 non-Bush controls are at a distance of 5.3 or less.

East Africa, North Africa, and North Europe

This set of (roughly) Mesolithic specimens was diagnosed by DISPOP as representing virtually anything but a population appropriate for modern inhabitants of their several areas.

To resume, the first four are late prehistoric East African (see Leakey 1935). For the Elmenteita pair, nearest neighbors (of which there are plenty within a limit of a 5.3 distance) represent peoples far away, with nothing that would suggest African affiliation in terms of the sub-Saharan samples in the universe. For *#1922*, Elmenteita B, the presence of 65 neighbors with a distance of 5.3 or less would suggest assignment among various non-African populations which would be at least as acceptable as the probabilities generated by DISPOP: Guam (P=.09), Hawaii (P=.03), etc.

Nakuru 9 and Willey's Kopje 3 give entirely similar results, seeming here a little more remote from the control skulls generally. The notable thing is the absence of suggestions of African affiliation. Remarks have been made above,

under DISPOP, as to the possible considerable importance of this evidence for late population history in East Africa.

Afalou 5 and 9 from Algeria have similarly protean affiliations, although those with Tasmanians, seen in DISPOP, are reduced. Afalou 9 here in fact has a good number of the Zalavár population as near neighbors, so that a basic European relationship is suggested, which is the first expectation from previous examinations of the Afalou skulls by others.

The situation is rather similar for actual European crania of the Mesolithic, the Teviec pair from Brittany. Teviec 11, *#2078*, reads as Polynesian, but Teviec 16 has Ainu and Zalavár as likely neighbors. Again, perhaps robusticity is involved in the nearness of Polynesians and Guam to Teviec 11, but this does not help in ethnic diagnosis.

All of this recalls a matter discussed under DISPOP, where both ethnic connections and general distance from moderns were problematic. Here, ethnic connections are similarly odd but a number of control skulls are within reasonable distance of the targets. Perhaps, as suggested earlier, the size factor has been damped by the use of C-scores, developed to remove size, while nevertheless robusticity affects shape in ways that distort shape relative to possible modern congeners.

This is where the work of Lahr is pertinent. As noted above, with her metric and phenetic assessment of the full North African sample she found that the robustness and shape of the crania set that sample outside her modern series, while also failing to indicate any clear relation to one of these modern regional groups. Clearly, it is to be expected that some such phylogenetic relation exists—in a word, from other indications, a connection with "Caucasoids"—but cannot be found in the matrix of information from modern skulls.

Upper Paleolithic

All four of these skulls have Europeans, or at least Egyptians, as nearest neighbors, most satisfactorily in the cases of Mladeč 1 and

TABLE 9C-1
POPKIN results: Kurgan #1559

```
Zalava   1    2    4    5    7    8   10   11   12   14   15   17   18   21   25    32   33   34   39   41   45   46   47   58   62   72   88  106  108  193    39
        2.2  3.3  3.5  3.6  3.6  3.7  4.0  4.0  4.0  4.1  4.1  4.2  4.2  4.2  4.3   4.5  4.5  4.5  4.6  4.6  4.7  4.7  4.7  4.9  5.0  5.1  5.2  5.4  5.4  5.9   4.4

Berg    20   28   31   44   48   53   64   65   66   71   74   75   92   95   99   112  115  118  134  141  203  205  218  227  247  261  280  322  363  470   145
        4.2  4.3  4.5  4.7  4.7  4.8  5.0  5.0  5.0  5.1  5.1  5.1  5.3  5.3  5.3   5.4  5.4  5.5  5.6  5.6  6.0  6.0  6.0  6.1  6.2  6.3  6.4  6.5  6.7  7.2   5.5

Peru     6   16   23   49   55   60   78   82   91   94  120  121  122  128  137   140  145  147  150  165  186  239  245  266  268  297  333  337  347  430   163
        3.6  4.2  4.2  4.7  4.8  4.9  5.1  5.2  5.3  5.3  5.5  5.5  5.5  5.5  5.6   5.6  5.6  5.7  5.7  5.8  5.9  6.2  6.2  6.3  6.3  6.4  6.6  6.6  6.7  7.0   5.6

Norse    9   13   22   24   30   36   40   43   56   81   96  103  110  117  124   125  138  149  158  170  194  195  206  207  256  318  359  408  525  597   164
        3.8  4.1  4.2  4.3  4.4  4.5  4.6  4.7  4.8  5.2  5.3  5.4  5.4  5.5  5.5   5.5  5.6  5.7  5.7  5.8  5.9  5.9  6.0  6.0  6.2  6.5  6.7  6.9  7.4  7.8   5.5

Egypt   19   27   37   50   52   54   57   87   89  105  123  157  179  188  189   211  223  225  237  255  267  320  325  336  352  384  391  448  549  621   219
        4.2  4.3  4.5  4.7  4.8  4.8  4.8  5.2  5.2  5.4  5.5  5.7  5.9  5.9  5.9   6.0  6.1  6.1  6.2  6.2  6.3  6.5  6.6  6.6  6.7  6.8  6.8  7.1  7.5  7.9   5.9

Philip  42   59   98  104  113  133  162  172  175  180  192  215  222  284  288   289  299  304  331  349  356  365  375  376  422  437  496  544  578  638   287
        4.6  4.9  5.3  5.4  5.4  5.6  5.8  5.8  5.8  5.9  5.9  6.0  6.1  6.4  6.4   6.4  6.4  6.5  6.6  6.7  6.7  6.7  6.8  6.8  6.9  7.0  7.3  7.5  7.7  8.0   6.3

Guam    51   67   69  153  156  160  161  164  178  199  226  235  258  274  278   287  298  301  305  323  329  372  428  429  455  473  480  491  587  674   295
        4.8  5.0  5.1  5.7  5.7  5.7  5.8  5.8  5.9  5.9  6.1  6.1  6.3  6.3  6.3   6.4  6.4  6.4  6.5  6.5  6.6  6.7  7.0  7.0  7.1  7.2  7.2  7.3  7.7  8.2   6.4

Atayal   3   35   86  109  130  142  166  190  196  204  208  212  230  241  257   270  275  300  360  388  395  413  425  449  464  477  577  646  733    .   296
        3.4  4.5  5.2  5.4  5.5  5.6  5.8  5.9  5.9  6.0  6.0  6.0  6.1  6.2  6.3   6.3  6.3  6.4  6.7  6.8  6.8  6.9  6.9  7.1  7.2  7.2  7.8  8.0  8.6    .   6.3

Ainu    26   93  129  148  151  159  198  202  221  228  254  262  271  279  281   303  324  342  353  378  387  398  409  426  443  458  494  515  529  533   306
        4.3  5.3  5.5  5.7  5.7  5.7  5.9  6.0  6.0  6.1  6.2  6.3  6.3  6.4  6.4   6.5  6.6  6.6  6.7  6.8  6.8  6.9  6.9  7.0  7.0  7.1  7.3  7.4  7.5  7.5   6.4

S.Cruz  38  114  131  144  146  167  173  174  201  216  248  251  264  285  307   311  312  334  340  424  434  435  439  466  468  471  512  513  518  551   313
        4.6  5.4  5.5  5.6  5.7  5.8  5.8  5.8  5.9  6.0  6.2  6.2  6.3  6.4  6.5   6.5  6.5  6.6  6.6  6.9  7.0  7.0  7.0  7.2  7.2  7.2  7.4  7.4  7.4  7.6   6.4

Tolai   61   83   84  116  127  132  139  152  155  171  177  191  210  217  259   282  292  293  309  326  474  490  553  590  605  616  632  641  652  720   325
        4.9  5.2  5.2  5.5  5.5  5.5  5.6  5.6  5.7  5.7  5.8  5.8  5.9  6.0  6.0  6.3   6.4  6.4  6.4  6.5  6.6  6.7  7.2  7.3  7.6  7.8  7.8  7.9  7.9  8.0  8.1  8.5   6.5

Hainan  63   90  101  163  176  181  187  209  220  234  236  244  291  343  348   355  389  400  411  415  420  452  453  501  530  541  613  647  658  681   352
        5.0  5.3  5.3  5.8  5.8  5.9  5.9  6.0  6.0  6.1  6.1  6.2  6.4  6.6  6.6   6.7  6.8  6.9  6.9  6.9  6.9  7.1  7.1  7.3  7.5  7.5  7.8  8.0  8.1  8.3   6.6

Austra  70   73   77  100  102  111  126  231  238  249  250  263  273  319  344   390  438  441  461  493  534  537  556  599  611  669  684  744  762  812   385
        5.1  5.1  5.1  5.3  5.3  5.4  5.5  6.1  6.2  6.2  6.2  6.3  6.3  6.5  6.6   6.8  7.0  7.0  7.1  7.3  7.5  7.5  7.6  7.8  7.8  8.2  8.3  8.7  8.9  9.6   6.8

Zulu    76   85  107  182  184  214  219  242  290  295  345  351  362  380  403   404  432  459  481  497  504  511  527  562  624  636  664  673  701  742   405
        5.1  5.2  5.4  5.9  5.9  6.0  6.0  6.2  6.4  6.4  6.6  6.7  6.7  6.8  6.9   6.9  7.0  7.1  7.2  7.3  7.3  7.4  7.4  7.6  7.9  8.0  8.2  8.2  8.4  8.7   6.9

NJapan 185  200  213  233  252  286  296  313  316  346  368  371  379  417  445   450  451  456  465  469  482  486  554  564  589  594  635  655  707  790   432
        5.9  5.9  6.0  6.1  6.2  6.4  6.4  6.5  6.5  6.6  6.7  6.7  6.8  6.9  7.0   7.1  7.1  7.1  7.2  7.2  7.2  7.3  7.6  7.6  7.8  7.8  8.0  8.1  8.4  9.2   7.0

Arikar  29  168  169  229  232  327  341  370  373  399  401  412  414  416  440   462  495  510  521  524  572  591  596  642  644  699  708  715  729  818   465
        4.4  5.8  5.8  6.1  6.1  6.6  6.6  6.7  6.8  6.9  6.9  6.9  6.9  6.9  7.0   7.1  7.3  7.4  7.4  7.4  7.7  7.8  7.8  8.0  8.0  8.4  8.4  8.5  8.6  9.7   7.2

Andama  97  224  253  272  308  328  338  354  364  386  394  397  407  442  457   488  492  499  500  503  531  574  600  637  643  650  662  688  713  789   466
        5.3  6.1  6.2  6.3  6.5  6.6  6.6  6.7  6.7  6.8  6.8  6.8  6.9  7.0  7.1   7.3  7.3  7.3  7.3  7.3  7.5  7.7  7.8  8.0  8.0  8.1  8.1  8.3  8.5  9.2   7.2

Tasman  68  265  276  277  283  294  335  374  419  433  447  472  483  514  522   539  543  545  550  552  592  633  634  657  682  703  710  737  766  779   503
        5.1  6.3  6.3  6.3  6.4  6.4  6.6  6.8  6.9  7.0  7.1  7.2  7.2  7.4  7.4   7.5  7.5  7.5  7.6  7.6  7.8  7.9  8.0  8.1  8.3  8.4  8.5  8.6  8.9  9.1   7.4

SJapan 119  260  310  321  330  357  405  427  431  479  485  508  509  517  520   548  561  569  585  588  619  623  628  630  680  695  732  738  748  810   524
        5.5  6.3  6.5  6.5  6.6  6.7  6.9  7.0  7.0  7.2  7.3  7.3  7.3  7.4  7.4   7.5  7.6  7.7  7.7  7.8  7.9  7.9  7.9  8.3  8.4  8.6  8.7  8.8  9.5   7.5

Teita  135  317  361  369  377  383  385  393  396  421  478  484  487  519  528   557  568  583  615  627  666  678  689  696  705  735  774  783  799  814   544
        5.6  6.5  6.5  6.7  6.8  6.8  6.8  6.8  6.8  6.9  7.2  7.2  7.3  7.4  7.5   7.6  7.7  7.7  7.9  7.9  8.2  8.2  8.3  8.4  8.4  8.6  9.0  9.1  9.3  9.6   7.6

Anyang  79   80  136  240  302  444  454  475  498  502  506  547  565  570  593   602  608  609  625  649  654  679  685  693  721  749  768  769  781  819   547
        5.2  5.2  5.6  6.2  6.4  7.0  7.1  7.2  7.3  7.3  7.3  7.5  7.7  7.7  7.8   7.8  7.8  7.8  7.9  8.0  8.1  8.2  8.3  8.4  8.5  8.8  9.0  9.0  9.1  9.7   7.6

Dogon  154  183  243  269  315  406  410  423  476  507  559  586  604  612  631   660  670  671  683  704  718  722  724  755  758  760  773  792  804  805   579
        5.7  5.9  6.2  6.3  6.5  6.9  6.9  6.9  7.2  7.3  7.6  7.7  7.8  7.8  7.9   8.1  8.2  8.2  8.3  8.4  8.5  8.5  8.6  8.8  8.9  8.9  9.0  9.2  9.5  9.5   7.8

Bushma 246  314  366  402  418  489  558  575  579  581  584  598  601  614  622   626  639  640  651  677  691  716  734  736  750  752  764  802  809  829   612
        6.2  6.5  6.7  6.9  6.9  7.3  7.6  7.7  7.7  7.7  7.7  7.8  7.8  7.8  7.9   7.9  8.0  8.0  8.1  8.2  8.4  8.5  8.6  8.6  8.8  8.8  8.9  9.4  9.5   10   8.0

Hawaii 339  358  446  463  505  536  542  546  576  580  607  610  617  620  659   663  686  687  694  698  714  723  727  740  743  765  784  797  820  825   636
        6.6  6.7  7.1  7.2  7.3  7.5  7.5  7.5  7.7  7.7  7.8  7.8  7.9  7.9  8.1   8.2  8.3  8.3  8.4  8.4  8.5  8.6  8.6  8.7  8.7  8.9  9.2  9.3  9.7  9.9   8.1

Eskimo 306  381  392  467  532  535  540  560  563  566  567  582  595  603  668   672  676  697  700  702  726  739  756  761  771  787  798  800  807  836   636
        6.5  6.8  6.8  7.2  7.5  7.5  7.5  7.6  7.6  7.7  7.7  7.7  7.8  7.8  8.1   8.2  8.2  8.4  8.4  8.4  8.6  8.7  8.8  8.9  9.0  9.2  9.3  9.4  9.5   11   8.2

Easter 350  367  460  538  555  571  573  618  629  645  648  661  667  675  692   706  712  719  731  745  778  780  782  786  794  811  816  817  821  822   676
        6.7  6.7  7.1  7.5  7.6  7.7  7.7  7.9  7.9  8.0  8.0  8.1  8.2  8.2  8.4   8.4  8.5  8.5  8.6  8.7  9.1  9.1  9.1  9.2  9.2  9.6  9.7  9.7  9.7  9.8   8.4

Buriat 143  197  332  526  606  653  656  711  725  730  741  747  753  757  763   770  772  776  777  791  803  808  813  823  827  830  835  837  838  839   706
        5.6  5.9  6.6  7.4  7.8  8.1  8.1  8.5  8.6  8.6  8.7  8.8  8.8  8.8  8.9   9.0  9.0  9.1  9.1  9.2  9.4  9.5  9.6  9.8   10   10   11   11   11   11   8.9

Morior 382  436  516  523  665  690  709  717  728  746  751  754  759  767  775   785  788  793  795  796  801  806  815  824  826  828  831  832  833  834   737
        6.8  7.0  7.4  7.4  8.2  8.4  8.4  8.5  8.6  8.8  8.8  8.8  8.9  9.0  9.1   9.2  9.2  9.2  9.3  9.3  9.4  9.5  9.6  9.8  9.9   10   10   10   10   11   9.0
```

TABLE 9C-2
POPKIN results: Kurgan #1558

```
Norse    1   2   4   7  11  14  16  17  20  22  23  26  27  36  37    38  39  41  45  47  51  52  58  62  65 117 129 147 162 711    68
        3.3 3.6 3.7 4.0 4.1 4.2 4.3 4.3 4.4 4.5 4.6 4.7 4.8 4.8        4.8 4.8 4.8 4.9 5.0 5.0 5.1 5.2 5.3 5.8 5.9 6.1 6.2 8.6        4.9

Zalava   3   5   6   8   9  10  12  15  18  31  40  46  48  49  53    54  68  70  71  93  94 110 115 136 140 156 158 258 260 435    86
        3.6 3.7 3.8 4.0 4.0 4.0 4.2 4.2 4.3 4.7 4.8 4.9 5.0 5.0 5.0    5.1 5.3 5.3 5.3 5.5 5.5 5.5 5.7 5.8 6.0 6.0 6.2 6.2 6.6 6.7 7.4   5.1

Egypt   19  21  25  28  35  43  55  60  63  64  77  85  90  92 101    102 113 134 142 143 150 152 154 168 171 174 227 276 280 353   120
        4.3 4.4 4.6 4.7 4.8 4.9 5.1 5.1 5.2 5.2 5.4 5.4 5.5 5.5 5.6    5.6 5.7 6.0 6.0 6.0 6.1 6.1 6.1 6.2 6.2 6.3 6.5 6.7 6.7 7.1   5.6

Berg    29  30  50  57  61  66  73  81  83  89  98  99 104 108 112    114 128 132 141 163 301 389 420 434 491 500 509 604 712 733   230
        4.7 4.7 5.0 5.1 5.2 5.3 5.4 5.4 5.4 5.5 5.6 5.6 5.7 5.7 5.7    5.7 5.9 6.0 6.0 6.2 6.9 7.2 7.4 7.4 7.6 7.6 7.7 8.1 8.6 8.7   6.2

Teita   13  44  69  72  75  88  91 124 126 130 137 155 159 170 205    213 233 245 250 257 265 284 287 291 331 417 418 467 594 781   233
        4.2 4.9 5.3 5.3 5.4 5.5 5.5 5.9 5.9 5.9 6.0 6.1 6.2 6.2 6.4    6.4 6.5 6.6 6.6 6.6 6.7 6.8 6.8 6.8 7.0 7.3 7.3 7.5 8.0 9.2   6.4

Ainu    56  59 100 107 118 123 131 182 203 218 220 234 237 252 290    329 333 342 380 415 432 478 506 521 533 595 611 621 658 732   334
        5.1 5.1 5.6 5.7 5.8 5.9 5.9 6.3 6.4 6.5 6.5 6.5 6.6 6.8        7.0 7.0 7.0 7.2 7.3 7.4 7.6 7.7 7.7 7.8 8.0 8.1 8.1 8.3 8.7   6.9

Anyang  34 111 125 133 144 165 176 201 222 249 253 283 309 327 332    346 350 352 356 390 411 423 448 471 513 530 593 596 620 685   338
        4.7 5.7 5.9 6.0 6.0 6.2 6.3 6.4 6.5 6.6 6.6 6.8 6.9 7.0 7.0    7.0 7.0 7.1 7.1 7.2 7.3 7.4 7.4 7.5 7.7 7.8 8.0 8.0 8.1 8.4   6.9

Guam    32 109 157 187 189 202 208 221 235 239 255 268 319 322 343    363 368 394 396 409 485 499 518 529 552 580 623 670 687 802   372
        4.7 5.7 6.2 6.3 6.3 6.4 6.4 6.5 6.5 6.6 6.6 6.7 6.9 7.0 7.0    7.1 7.1 7.3 7.3 7.3 7.6 7.6 7.7 7.7 7.8 7.9 8.1 8.3 8.4 9.5   7.1

Hainan 120 121 139 149 169 198 206 207 209 228 240 247 271 288 387    395 413 431 449 461 473 510 535 538 562 585 637 640 644 731   373
        5.8 5.8 6.0 6.1 6.2 6.4 6.4 6.4 6.4 6.5 6.6 6.6 6.7 6.8 7.2    7.3 7.3 7.4 7.4 7.5 7.5 7.7 7.8 7.8 7.9 8.0 8.2 8.2 8.2 8.7   7.1

Peru   164 180 184 190 194 210 225 236 278 318 324 328 334 335 337    376 382 397 422 428 438 463 480 599 630 647 664 665 688 691   394
        6.2 6.3 6.3 6.3 6.4 6.4 6.5 6.5 6.7 6.9 7.0 7.0 7.0 7.0 7.0    7.2 7.2 7.3 7.4 7.4 7.4 7.5 7.6 8.0 8.2 8.2 8.3 8.3 8.4 8.4   7.2

Philip  74 106 166 211 212 216 259 261 273 282 292 366 385 392 399    410 433 459 462 468 502 505 519 540 550 554 563 576 701 703   395
        5.4 5.7 6.2 6.4 6.4 6.5 6.7 6.7 6.8 6.8 7.1 7.2 7.2 7.3        7.3 7.4 7.5 7.5 7.6 7.7 7.7 7.8 7.8 7.9 7.9 8.5 8.6           7.2

SJapan  80  86 135 188 226 230 242 264 274 285 307 310 312 313 349    361 374 437 456 492 501 504 507 515 555 561 679 741 776 813   396
        5.4 5.5 6.0 6.3 6.5 6.5 6.6 6.6 6.7 6.7 6.8 6.9 6.9 6.9 7.0    7.1 7.2 7.4 7.5 7.6 7.6 7.6 7.7 7.7 7.7 7.8 7.9 8.4 8.9 9.1 9.7 7.2

Atayal  78  87 175 191 219 243 303 311 321 323 330 341 354 364 371    375 404 412 426 429 446 475 476 589 667 714 724 743 787   .   403
        5.4 5.5 5.6 3 6.3 6.5 6.6 6.9 6.9 6.9 7.0 7.0 7.0 7.1 7.1 7.1  7.2 7.3 7.3 7.4 7.4 7.4 7.5 7.6 8.0 8.3 8.6 8.7 8.8 9.3   .   7.3

Eskimo  04  7( 70 16[ 107 0T0 0[1 00T T[[ 770 T01 T00 070 0[0 0[7     0[0 070 077 000 00[ [1( [7( [01 [0( [17 [77 [07 [[0 770 7[[   076
        4.6 5.4 5.4 6.0 6.4 6.5 6.6 6.8 7.1 7.2 7.2 7.3 7.4 7.5 7.5    7.5 7.5 7.6 7.6 7.6 7.7 7.8 7.8 8.0 8.1 8.2 8.2 8.3 8.7 9.0   7.3

NJapan 160 161 172 193 199 204 214 238 262 263 272 362 393 442 450    474 496 503 546 569 570 605 645 654 657 684 709 728 748 791   445
        6.2 6.2 6.2 6.4 6.4 6.4 6.5 6.5 6.5 6.7 6.7 7.1 7.2 7.4 7.5    7.5 7.6 7.6 7.8 7.9 7.9 8.1 8.2 8.3 8.3 8.4 8.6 8.7 8.9 9.4   7.4

Arikar  97 122 127 173 178 244 267 279 286 325 338 357 365 487 489    514 531 551 568 571 600 602 656 676 678 696 698 704 716 735   454
        5.6 5.8 5.9 6.3 6.3 6.6 6.7 6.8 7.0 7.0 7.1 7.1 7.6 7.6        7.7 7.7 7.8 7.9 8.1 8.1 8.3 8.4 8.4 8.5 8.5 8.6 8.6 8.7       7.4

Bushma  33  96 186 196 200 223 306 347 348 370 372 388 405 406 469    481 497 526 547 582 584 610 616 632 677 680 683 690 785 798   458
        4.7 5.5 6.3 6.4 6.4 6.5 6.9 7.0 7.0 7.1 7.1 7.2 7.3 7.3 7.5    7.6 7.6 7.7 7.8 7.9 8.0 8.1 8.1 8.2 8.4 8.4 8.4 8.4 9.3 9.5   7.5

S.Cruz 151 167 183 224 246 277 281 320 351 360 367 386 407 419 443    484 534 574 577 592 606 615 634 682 700 706 723 730 789 800   478
        6.1 6.2 6.3 6.5 6.6 6.7 6.7 6.9 7.0 7.1 7.1 7.2 7.3 7.3 7.4    7.6 7.8 7.9 7.9 8.0 8.1 8.1 8.2 8.4 8.5 8.6 8.7 8.7 9.4 9.5   7.6

Tolai  148 185 195 229 241 256 294 298 369 391 416 430 451 458 483    486 490 493 520 537 559 653 674 695 697 705 759 783 792 817   483
        6.1 6.3 6.4 6.5 6.6 6.6 6.8 6.8 7.1 7.2 7.3 7.4 7.5 7.5 7.6    7.6 7.6 7.6 7.7 7.8 7.8 8.3 8.3 8.5 8.5 8.6 8.9 9.3 9.4 9.9   7.6

Andama  84 138 266 270 296 297 314 315 339 358 401 403 472 508 539    557 558 567 587 588 590 618 626 629 669 694 699 749 760 812   490
        5.4 6.0 6.7 6.7 6.8 6.8 6.9 6.9 7.0 7.1 7.3 7.3 7.5 7.7 7.8    7.8 7.8 7.9 8.0 8.0 8.0 8.1 8.1 8.1 8.3 8.5 8.5 8.8 8.9 9.7   7.6

Easter  82 116 181 192 248 275 336 340 402 408 436 464 482 494 498    542 556 619 625 627 631 663 675 681 689 725 726 761 763 772   500
        5.4 5.8 6.3 6.3 6.6 6.7 7.0 7.0 7.3 7.3 7.4 7.5 7.6 7.6 7.6    7.8 7.8 8.1 8.1 8.1 8.2 8.3 8.3 8.4 8.4 8.7 8.7 8.9 9.0 9.1   7.7

Zulu    67  95 103 215 269 299 308 377 445 465 479 523 524 525 532    548 549 566 578 591 603 607 643 651 655 662 758 782 805 828   502
        5.3 5.5 5.6 6.5 6.7 6.9 6.9 7.2 7.4 7.5 7.6 7.7 7.7 7.7 7.8    7.8 7.8 7.9 7.9 8.0 8.1 8.1 8.2 8.2 8.3 8.3 8.9 9.3 9.6  10    7.7

Morior 146 153 177 231 295 300 316 383 427 466 511 527 543 545 575    608 638 639 659 672 686 713 718 721 729 739 769 771 775 786   541
        6.0 6.1 6.3 6.5 6.8 6.9 6.9 7.2 7.4 7.5 7.7 7.7 7.8 7.8 7.9    8.1 8.2 8.2 8.3 8.3 8.4 8.6 8.6 8.7 8.7 8.8 9.0 9.1 9.1 9.3   7.9

Buriat  42 179 217 305 317 345 359 512 553 609 617 646 648 668 692    707 727 738 746 756 767 795 796 803 804 811 829 835 836 837   617
        4.9 6.3 6.5 6.9 6.9 7.0 7.1 7.7 7.8 8.1 8.1 8.2 8.2 8.3 8.4    8.6 8.7 8.8 8.8 8.9 9.0 9.4 9.4 9.5 9.5 9.7  10  10  11  11   8.4

Hawaii 302 304 326 400 421 424 444 560 565 572 581 601 614 628 650    666 673 720 722 742 745 747 757 778 788 794 808 810 819 824   626
        6.9 6.9 7.0 7.3 7.4 7.4 7.4 7.8 7.9 7.9 7.9 8.1 8.1 8.1 8.2    8.3 8.3 8.7 8.7 8.8 8.8 8.8 8.9 9.2 9.4 9.4 9.7 9.7  10  10   8.4

Tasman 105 373 379 414 425 447 453 528 544 624 649 652 693 702 719    750 754 755 762 764 766 770 773 780 815 821 822 826 833 838   651
        5.7 7.1 7.2 7.3 7.4 7.4 7.5 7.7 7.8 8.1 8.2 8.2 8.4 8.5 8.7    8.8 8.9 8.9 8.9 9.0 9.0 9.1 9.1 9.2 9.8  10  10  10  10  11   8.6

Austra 119 289 344 517 522 566 573 598 622 635 636 641 661 671 717    734 737 740 744 752 768 779 784 801 806 809 818 830 831 839   663
        5.8 6.8 7.0 7.7 7.7 7.9 7.9 8.0 8.1 8.2 8.2 8.2 8.3 8.3 8.6    8.7 8.8 8.8 8.8 8.9 9.0 9.2 9.3 9.5 9.6 9.7 9.9  10  10  11   8.6

Dogon  254 384 440 441 452 455 579 583 597 612 708 710 715 751 753    774 777 790 793 797 799 807 814 816 820 823 825 827 832 834   685
        6.6 7.2 7.4 7.4 7.5 7.5 7.9 8.0 8.0 8.1 8.6 8.6 8.6 8.9 8.9    9.1 9.1 9.4 9.4 9.4 9.5 9.6 9.8 9.8  10  10  10  10  10  10   8.8
```

TABLE 9C-3
POPKIN results: Natufian #1775

```
Zalava    4    5    7    9   15   20   23   26   35   36   44   51   53   61   62      75   79   80   90  101  104  125  134  165  196  219  224  295  420  440    107
        3.9  4.2  4.4  4.4  4.6  4.7  4.7  4.8  5.0  5.0  5.0  5.1  5.1  5.2  5.2     5.3  5.3  5.4  5.4  5.5  5.5  5.7  5.7  5.9  6.1  6.2  6.2  6.5  6.9  7.0    5.3

Ainu      3    6    8   29   66   94   95   97  133  177  179  186  188  208  216     222  228  254  269  289  290  319  338  376  382  388  395  408  524  707    236
        3.6  4.3  4.4  4.8  5.2  5.4  5.4  5.5  5.7  6.0  6.0  6.0  6.0  6.1  6.1     6.2  6.2  6.3  6.4  6.5  6.5  6.6  6.7  6.8  6.8  6.9  6.9  7.3  8.3    6.1

Philip   17   22   31   39   49   60   73   82  105  131  139  147  161  207  209     223  240  262  288  322  358  372  417  445  477  480  489  514  535  717    254
        4.6  4.7  4.8  5.0  5.1  5.2  5.3  5.4  5.5  5.7  5.7  5.8  5.9  6.1  6.1     6.2  6.3  6.4  6.5  6.6  6.8  6.8  6.9  7.0  7.2  7.2  7.2  7.3  7.4  8.4    6.2

Zulu      1   27   48   67   83   91   92  103  114  143  153  184  191  212  238     239  243  251  304  326  340  350  351  370  374  421  540  551  569  690    256
        3.5  4.8  5.1  5.3  5.4  5.4  5.4  5.5  5.6  5.8  5.8  6.0  6.0  6.1  6.3     6.3  6.3  6.3  6.6  6.6  6.6  6.7  6.7  6.8  6.8  6.8  6.9  7.4  7.4  7.5  8.2   6.2

Tasman    2   10   11   32   55   81   86   99  106  116  129  132  148  150  159     203  233  247  305  363  402  416  419  437  459  464  541  734  747  774    269
        3.5  4.5  4.5  4.9  5.2  5.4  5.4  5.5  5.5  5.6  5.7  5.7  5.8  5.8  5.9     6.1  6.3  6.3  6.6  6.8  6.9  6.9  6.9  7.0  7.1  7.1  7.4  8.6  8.7  9.0    6.2

Austra   12   13   24   28   37   43   56   68   77   78  128  137  194  211  249     267  284  286  342  357  362  400  407  415  458  525  627  658  711  778    274
        4.5  4.6  4.7  4.8  5.0  5.0  5.2  5.3  5.3  5.3  5.7  5.7  6.1  6.1  6.3     6.4  6.5  6.5  6.7  6.8  6.8  6.9  6.9  6.9  7.1  7.3  7.8  8.0  8.4  9.0    6.3

Berg     14   21   38   41   46   54   70   93  102  108  112  138  151  230  248     273  291  299  312  315  404  411  476  479  487  500  632  638  686  706    279
        4.6  4.7  5.0  5.0  5.1  5.2  5.3  5.4  5.5  5.5  5.5  5.7  5.8  6.2  6.3     6.4  6.5  6.5  6.6  6.6  6.9  6.9  7.2  7.2  7.2  7.2  7.8  7.9  8.2  8.3    6.3

Atayal   34   42   59   96  100  107  110  160  170  172  180  195  227  231  244     265  278  297  302  389  414  430  433  451  478  507  511  586  656    .   280
        4.9  5.0  5.2  5.5  5.5  5.5  5.5  5.9  5.9  5.9  6.0  6.1  6.2  6.2  6.3     6.4  6.4  6.5  6.6  6.9  6.9  6.9  7.0  7.0  7.2  7.3  7.3  7.6  8.0    .   6.3

S.Cruz   25   50  109  122  130  140  141  164  182  218  235  237  255  256  259     311  325  336  346  354  365  406  467  492  515  520  528  538  629  633    310
        4.8  5.1  5.5  5.7  5.7  5.8  5.8  5.9  6.0  6.2  6.3  6.3  6.4  6.4  6.4     6.6  6.6  6.7  6.7  6.8  6.8  6.9  7.1  7.2  7.3  7.3  7.3  7.4  7.8  7.9    6.5

Norse    16   33   40   57   72   85  118  119  127  135  157  200  236  242  246     306  335  390  394  436  443  462  543  556  564  567  579  635  646  675    314
        4.6  4.9  5.0  5.2  5.3  5.4  5.6  5.6  5.7  5.7  5.9  6.1  6.3  6.3  6.3     6.6  6.7  6.9  6.9  7.0  7.0  7.1  7.4  7.5  7.5  7.5  7.6  7.9  7.9  8.1    6.4

Tolai    18   19   52   64   69  142  156  158  187  189  204  214  226  257  276     282  307  314  316  465  484  501  517  565  570  612  613  621  666  737    327
        4.7  4.7  5.1  5.2  5.3  5.8  5.9  5.9  6.0  6.0  6.1  6.1  6.2  6.4  6.4     6.5  6.6  6.6  6.6  7.1  7.2  7.2  7.3  7.5  7.5  7.8  7.8  7.8  8.1  8.6    6.5

Egypt   121  126  154  169  174  175  176  192  197  205  213  241  281  344  345     349  359  368  371  401  403  424  427  491  537  558  559  587  634  644    341
        5.6  5.7  5.9  6.0  6.0  6.0  6.0  6.1  6.1  6.3  6.5  6.7  6.7     6.8  6.8  6.9  6.9  7.2  7.4  7.5  7.5  7.7  7.9  7.9    6.6

Peru     47   58  144  146  162  193  206  220  260  270  303  324  329  347  355     356  413  423  438  442  456  473  498  504  506  522  546  602  606  741    363
        5.1  5.2  5.8  5.8  5.9  6.1  6.1  6.2  6.4  6.4  6.6  6.6  6.6  6.7  6.8     6.8  6.9  6.9  7.0  7.0  7.0  7.1  7.2  7.2  7.3  7.3  7.4  7.7  7.8  8.6    6.7

NJapan   65   84   87  111  123  152  181  215  221  250  279  292  296  332  352     385  405  418  429  516  531  532  533  539  568  597  614  615  653  678    368
        5.2  5.4  5.4  5.5  5.7  5.8  6.0  6.1  6.2  6.3  6.5  6.5  6.5  6.7  6.8     6.9  6.9  6.9  6.9  7.3  7.4  7.4  7.4  7.4  7.5  7.7  7.8  7.8  8.0  8.1    6.7

Arikar   74  113  120  124  145  171  301  309  328  330  348  373  383  386  393     422  425  448  455  483  490  499  583  589  611  642  652  701  715  771    413
        5.3  5.5  5.6  5.7  5.8  5.9  6.6  6.6  6.6  6.6  6.7  6.7  6.8  6.8  6.9  6.9     6.9  6.9  7.0  7.0  7.2  7.2  7.2  7.6  7.6  7.8  7.9  8.0  8.3  8.4  8.9    6.9

Hainan   76  115  155  183  185  261  275  277  313  337  339  384  432  435  441     452  472  474  509  526  530  563  628  640  651  655  669  679  714  796    439
        5.3  5.6  5.9  6.0  6.0  6.4  6.4  6.4  6.6  6.7  6.7  6.8  7.0  7.0  7.0     7.0  7.1  7.1  7.3  7.3  7.3  7.5  7.8  7.9  8.0  8.0  8.1  8.1  8.4  9.3    7.1

Guam    163  167  225  232  268  274  294  298  317  320  366  367  381  397  399     426  434  495  505  544  555  560  600  671  693  729  732  746  765  768    456
        5.9  5.9  6.2  6.3  6.4  6.4  6.5  6.5  6.6  6.6  6.8  6.8  6.8  6.9  6.9     6.9  7.0  7.2  7.3  7.4  7.5  7.5  7.7  8.1  8.2  8.5  8.6  8.7  8.9  8.9    7.2

SJapan   71  173  253  283  285  287  318  341  361  409  446  450  466  468  481     496  510  549  562  580  609  623  626  636  641  665  700  721  745  821    486
        5.3  6.0  6.3  6.5  6.5  6.5  6.6  6.7  6.8  6.9  7.0  7.0  7.1  7.1  7.2     7.2  7.3  7.4  7.5  7.6  7.7  7.8  7.8  7.9  7.9  8.1  8.3  8.5  8.7   10    7.3

Andama   30   88   98  168  234  327  333  353  379  392  396  449  494  512  542     548  550  554  582  593  650  667  677  697  713  716  719  724  743  781    490
        4.8  5.4  5.5  5.9  6.3  6.6  6.7  6.8  6.8  6.9  6.9  7.0  7.2  7.3  7.4     7.4  7.4  7.5  7.6  7.7  8.0  8.1  8.1  8.2  8.4  8.4  8.4  8.5  8.6  9.0    7.3

Teita    89  190  210  252  264  272  280  343  369  377  391  410  412  428  469     486  523  591  601  608  620  643  680  698  699  708  749  773  793  818    492
        5.4  6.0  6.1  6.3  6.4  6.4  6.5  6.7  6.8  6.8  6.9  6.9  6.9  6.9  7.1     7.2  7.3  7.6  7.7  7.7  7.8  7.9  8.1  8.3  8.3  8.3  8.7  8.9  9.2   10    7.4

Easter  178  201  229  245  263  266  271  380  387  398  460  475  488  493  552     553  571  575  585  590  637  659  660  674  676  681  683  731  755  762    503
        6.0  6.1  6.2  6.3  6.4  6.4  6.4  6.8  6.9  6.9  7.1  7.1  7.2  7.2  7.5     7.5  7.5  7.6  7.6  7.6  7.9  8.0  8.0  8.1  8.1  8.2  8.2  8.5  8.8  8.8    7.4

Anyang   45   63  136  293  323  331  364  431  497  503  508  513  518  534  557     572  576  594  630  657  664  685  687  704  720  723  748  751  764  790    529
        5.0  5.2  5.7  6.5  6.6  6.6  6.8  6.9  7.2  7.2  7.3  7.3  7.3  7.4  7.5     7.5  7.6  7.7  7.8  8.0  8.0  8.2  8.3  8.4  8.5  8.7  8.7  8.8  8.9  9.2    7.5

Bushma  149  202  217  308  310  334  378  439  453  454  463  527  545  566  596     604  618  647  661  682  689  722  725  726  750  757  766  777  795  816    556
        5.8  6.1  6.2  6.6  6.6  6.7  6.8  7.0  7.0  7.0  7.1  7.3  7.4  7.5  7.7     7.7  7.8  8.0  8.0  8.2  8.2  8.5  8.5  8.5  8.7  8.8  8.9  9.0  9.3   10    7.7

Morior  117  166  199  457  470  482  502  536  573  577  588  592  599  619  625     639  649  662  668  670  673  692  695  696  705  727  752  753  761  769    587
        5.6  5.9  6.1  7.1  7.1  7.2  7.2  7.4  7.5  7.6  7.6  7.7  7.7  7.8  7.8     7.9  8.0  8.0  8.1  8.1  8.1  8.2  8.2  8.2  8.3  8.5  8.7  8.7  8.8  8.9    7.7

Hawaii  258  300  321  360  375  444  461  471  561  574  598  603  605  607  616     622  645  654  672  684  710  712  739  756  763  767  776  779  785  786    600
        6.4  6.5  6.6  6.8  6.8  7.0  7.1  7.1  7.5  7.5  7.7  7.7  7.7  7.7  7.8     7.8  7.9  8.0  8.1  8.2  8.4  8.4  8.6  8.8  8.8  8.9  9.0  9.0  9.1  9.1    7.9

Dogon   198  447  485  519  529  547  578  584  595  610  631  648  691  702  709     718  728  740  744  754  758  770  787  788  797  812  813  817  819  829    672
        6.1  7.0  7.2  7.3  7.3  7.4  7.6  7.6  7.7  7.8  7.8  8.0  8.2  8.3  8.3     8.4  8.5  8.6  8.7  8.7  8.8  8.9  9.1  9.1  9.4   10   10   10   10   11    8.4

Eskimo  521  581  663  688  694  703  730  733  735  736  738  759  760  772  780     789  791  792  794  798  799  802  804  805  806  814  815  820  822  833    756
        7.3  7.6  8.0  8.2  8.2  8.3  8.5  8.6  8.6  8.6  8.8  8.8  8.9  9.0  9.0     9.2  9.2  9.2  9.2  9.4  9.5  9.8  9.9  9.9  9.9   10   10   10   10   11    9.1

Buriat  617  624  742  775  782  783  784  800  801  803  807  808  809  810  811     823  824  825  826  827  828  830  831  832  834  835  836  837  838  839    801
        7.8  7.8  8.6  9.0  9.1  9.1  9.1  9.5  9.6  9.8  9.9  9.9   10   10   10      10   10   11   11   11   11   11   11   11   11   12   12   12   12   13   10.3
```

Table 9C-4
POPKIN results: Fish Hoek #1703

```
Bushma   1    2    3    4    5    6    8    9   10   11   12   13   15   17   19   20   24   27   30   32   35   37   38   40   43   45   60   63   81  116     28
        3.8  4.0  4.3  4.5  4.5  4.6  4.7  4.7  4.7  4.8  5.0  5.0  5.2  5.2  5.4  5.4  5.5  5.6  5.9  5.9  6.0  6.1  6.2  6.2  6.3  6.3  6.7  6.7  7.0  7.3    5.5

S.Cruz  18   25   26   58   59   68   70   86   97   98  102  107  115  121  135  140  143  154  156  179  190  218  263  267  309  350  372  454  478  488    178
        5.3  5.6  5.6  6.7  6.7  6.8  6.8  7.0  7.1  7.1  7.2  7.2  7.3  7.4  7.5  7.5  7.6  7.6  7.6  7.7  7.8  7.9  8.1  8.2  8.4  8.5  8.6  9.0  9.1  9.1    7.5

Zalava  29   33   34   42   44   53   55   56   65   76   92   99  117  132  139  141  158  192  197  200  207  223  237  247  256  305  322  378  509  728    179
        5.8  6.0  6.0  6.3  6.3  6.6  6.6  6.6  6.7  6.9  7.1  7.1  7.3  7.4  7.5  7.5  7.6  7.8  7.8  7.8  7.8  7.9  8.0  8.0  8.1  8.1  8.3  8.4  8.6  9.2   10     7.5

Norse   16   48   61   64   89   90   94  100  104  106  109  122  137  145  193  194  195  233  240  248  252  257  258  266  288  299  341  380  425  484    195
        5.2  6.5  6.7  6.7  7.0  7.0  7.1  7.1  7.2  7.2  7.3  7.4  7.5  7.6  7.8  7.8  7.8  8.0  8.1  8.1  8.1  8.1  8.1  8.2  8.3  8.3  8.5  8.7  8.9  9.1    7.6

Berg    21   22   23   36   39   46   50   51   80   82   85  146  151  166  181  199  203  205  208  219  221  319  327  382  475  523  561  632  657  683    233
        5.5  5.5  5.5  6.1  6.2  6.4  6.5  6.5  7.0  7.0  7.0  7.6  7.6  7.7  7.8  7.8  7.9  7.9  7.9  7.9  8.0  8.4  8.5  8.7  9.1  9.3  9.5  9.8   10   10     7.7

Austra  52   77   87   93  110  111  120  130  142  171  177  182  212  213  226  259  265  281  302  344  376  391  402  440  467  469  550  596  600  691    285
        6.6  6.9  7.0  7.1  7.3  7.3  7.4  7.4  7.6  7.7  7.7  7.8  7.9  7.9  8.0  8.1  8.1  8.2  8.3  8.5  8.6  8.7  8.8  8.9  9.1  9.1  9.4  9.6  9.6   10     8.2

Teita   54   57   66   72   78   88   95  105  159  162  176  211  229  236  242  254  255  274  303  349  351  366  397  404  472  547  551  652  678  696    286
        6.6  6.6  6.8  6.8  6.9  7.0  7.1  7.2  7.6  7.7  7.7  7.9  8.0  8.0  8.1  8.1  8.1  8.2  8.3  8.5  8.5  8.6  8.7  8.8  9.1  9.4  9.4  9.9   10   10     8.1

Ainu    28   84  119  123  124  129  172  173  202  204  217  244  280  292  306  313  324  360  361  368  379  412  416  423  455  473  503  638  664  765    318
        5.7  7.0  7.3  7.4  7.4  7.4  7.7  7.7  7.8  7.9  7.9  8.1  8.2  8.3  8.3  8.4  8.4  8.6  8.6  8.6  8.6  8.8  8.9  8.9  9.0  9.1  9.2  9.8   10   11     8.3

Egypt   31  133  138  149  163  183  187  198  224  239  246  269  287  289  308  316  326  340  374  394  400  409  414  442  456  487  528  533  556  576    320
        5.9  7.5  7.5  7.6  7.7  7.8  7.8  7.8  8.0  8.1  8.1  8.2  8.2  8.3  8.4  8.4  8.4  8.5  8.6  8.7  8.8  8.8  8.9  9.0  9.0  9.1  9.3  9.3  9.4  9.5    8.3

Zulu    14   41   74   79  103  114  131  148  232  234  268  278  282  296  298  318  325  329  388  401  421  426  466  494  526  564  609  635  649  676    327
        5.1  6.2  6.8  7.0  7.2  7.3  7.4  7.6  8.0  8.0  8.2  8.2  8.2  8.3  8.3  8.4  8.4  8.5  8.7  8.8  8.9  8.9  9.1  9.2  9.3  9.5  9.7  9.8  9.9   10     8.3

Andama   7   49   69  112  126  136  153  167  168  178  188  220  253  301  336  337  345  383  384  439  498  512  516  540  593  623  666  674  713  758    348
        4.7  6.5  6.8  7.3  7.4  7.5  7.6  7.7  7.7  7.7  7.7  7.8  7.9  8.1  8.3  8.5  8.5  8.7  8.7  8.9  9.2  9.2  9.3  9.3  9.6  9.8   10   10   10   11     8.4

Philip  96  108  144  180  184  225  238  249  295  304  310  311  314  328  338  348  354  355  356  358  370  437  452  491  535  568  581  588  636  637    356
        7.1  7.3  7.6  7.7  7.8  8.0  8.1  8.1  8.3  8.3  8.4  8.4  8.4  8.5  8.5  8.5  8.5  8.5  8.5  8.5  8.6  8.9  9.0  9.2  9.3  9.5  9.6  9.6  9.8  9.8    8.5

NJapan  62   75  125  127  150  206  210  215  243  251  260  273  291  297  312  381  385  422  436  438  445  463  476  537  607  654  686  692  712  738    372
        6.7  6.9  7.4  7.4  7.6  7.9  7.9  7.9  8.0  8.0  8.0  8.0  8.0  8.3  8.3  8.7  8.7  8.9  8.9  8.9  9.0  9.1  9.1  9.3  9.7  9.9   10   10   10   10     8.6

Atayal  67  134  209  241  245  261  279  321  330  335  347  353  362  390  424  428  446  480  500  525  530  531  544  602  625  629  653  672  679    .    419
        6.8  7.5  7.9  8.1  8.1  8.1  8.2  8.4  8.5  8.5  8.5  8.5  8.5  8.6  8.7  8.9  8.9  9.0  9.1  9.2  9.3  9.3  9.3  9.3  9.6  9.8  9.8  9.9   10   10    .     8.8

Tasman 113  118  128  152  160  169  222  230  272  277  283  284  357  367  375  447  448  482  520  583  603  616  669  687  705  726  735  739  779  801    435
        7.3  7.3  7.4  7.6  7.6  7.7  8.0  8.0  8.2  8.2  8.2  8.2  8.5  8.6  8.6  9.0  9.0  9.1  9.3  9.6  9.6  9.7   10   10   10   10   10   10   11   11     9.0

Peru   161  170  175  196  201  270  276  290  294  323  333  399  406  413  461  490  499  501  502  514  515  536  613  614  619  630  631  650  695  778    439
        7.6  7.7  7.7  7.8  7.8  8.2  8.3  8.3  8.4  8.5  8.7  8.8  8.8  8.9  9.1  9.2  9.2  9.2  9.2  9.2  9.3  9.7  9.7  9.7  9.8  9.8  9.9  9.9   10   11     8.9

Tolai  164  165  191  216  262  317  343  363  369  410  427  433  441  444  450  481  486  493  519  521  541  612  615  621  626  670  694  729  748  770    471
        7.7  7.7  7.8  7.9  8.1  8.4  8.5  8.6  8.6  8.8  8.9  8.9  9.0  9.0  9.0  9.1  9.1  9.2  9.3  9.3  9.3  9.7  9.7  9.7  9.8   10   10   11   11     9.1

Anyang  47   83  228  231  286  359  371  392  407  418  451  459  462  504  513  532  557  562  567  569  572  574  585  624  639  648  658  715  766  803    486
        6.4  7.0  8.0  8.0  8.2  8.5  8.6  8.7  8.8  8.9  9.0  9.1  9.1  9.2  9.2  9.3  9.4  9.5  9.5  9.5  9.5  9.5  9.6  9.8  9.8  9.9   10   10   11   11     9.2

Hainan  91  155  250  300  386  387  389  395  411  417  434  460  470  511  517  524  538  542  545  546  560  608  677  707  716  718  724  750  753  771    508
        7.1  7.6  8.1  8.3  8.7  8.7  8.7  8.7  8.8  8.9  8.9  9.1  9.1  9.2  9.3  9.3  9.3  9.3  9.3  9.4  9.5  9.7   10   10   10   10   10   11   11   11     9.3

SJapan 147  174  189  339  346  420  429  432  457  468  483  492  495  505  508  529  553  566  584  587  606  627  640  645  660  662  667  706  768  824    517
        7.6  7.7  7.8  8.5  8.5  8.9  8.9  8.9  9.0  9.1  9.1  9.2  9.2  9.2  9.2  9.3  9.4  9.5  9.6  9.6  9.7  9.8  9.8  9.9   10   10   10   10   11   12     9.3

Dogon   73  101  271  275  332  365  396  415  430  464  465  496  506  522  552  570  571  589  618  641  675  690  719  725  733  742  782  789  793  794    536
        6.8  7.2  8.2  8.2  8.5  8.6  8.7  8.9  8.9  9.1  9.1  9.2  9.2  9.3  9.4  9.5  9.5  9.6  9.7  9.9   10   10   10   10   10   10   11   11   11   11     9.4

Arikar  71  214  293  315  377  393  403  431  435  485  510  558  575  591  594  610  620  633  655  685  697  701  704  711  721  722  781  786  795  797    562
        6.8  7.9  8.3  8.4  8.6  8.7  8.8  8.9  8.9  9.1  9.2  9.4  9.5  9.6  9.6  9.7  9.7  9.8  9.9   10   10   10   10   10   10   10   11   11   11   11     9.6

Guam   227  235  264  352  408  419  453  471  477  507  527  539  555  577  580  590  622  634  644  651  659  682  703  708  744  751  754  780  785  828    571
        8.0  8.0  8.1  8.5  8.8  8.9  9.0  9.1  9.1  9.2  9.3  9.3  9.4  9.5  9.6  9.6  9.8  9.8  9.9  9.9   10   10   10   10   11   11   11   11   11   12     9.6

Morior 285  320  334  390  405  449  479  548  554  559  573  579  582  598  599  643  665  671  684  688  693  709  710  732  737  743  745  761  783  812    601
        8.2  8.4  8.5  8.7  8.8  9.0  9.1  9.4  9.4  9.4  9.5  9.6  9.6  9.6  9.6  9.9   10   10   10   10   10   10   10   10   10   11   11   11   11     9.8

Buriat 186  307  331  342  373  443  518  578  597  604  605  680  698  700  720  723  731  734  740  749  762  772  773  805  807  808  810  815  826  838    646
        7.8  8.4  8.5  8.5  8.6  9.0  9.3  9.6  9.6  9.7  9.7   10   10   10   10   10   10   10   10   11   11   11   11   11   11   11   11   12   12   13   10.2

Eskimo 157  185  364  489  497  543  563  565  586  595  611  617  628  663  673  702  717  730  752  759  776  777  788  791  798  806  811  814  827  839    647
        7.6  7.8  8.6  9.2  9.2  9.3  9.5  9.5  9.6  9.6  9.7  9.7  9.8   10   10   10   10   10   11   11   11   11   11   11   11   11   11   12   14   10.2

Hawaii 458  474  592  601  642  647  661  668  757  760  763  764  767  769  775  784  790  796  799  804  813  817  820  821  829  831  834  835  836  837    745
        9.1  9.1  9.6  9.6  9.9  9.9   10   10   11   11   11   11   11   11   11   11   11   11   12   12   12   12   12   12   12   12   12   12   12   12   10.9

Easter 534  549  646  656  681  689  699  714  727  736  741  746  747  755  756  774  787  792  800  802  809  816  818  819  822  823  825  830  832  833    752
        9.3  9.4  9.9  9.9   10   10   10   10   10   10   10   10   11   11   11   11   11   11   11   11   11   12   12   12   12   12   12   12   12   12   10.9
```

TABLE 9C-5

POPKIN results: Elmenteita A #1921

```
Austra    3    4   10   12   14   15   26   27   30   38   48   61   69   85   90     95   98  130  133  138  143  180  199  240  300  317  333  360  399  497   136
         4.1  4.3  4.8  4.9  5.1  5.1  5.4  5.4  5.5  5.6  5.7  5.9  5.9  6.1  6.1    6.2  6.2  6.4  6.4  6.4  6.4  6.6  6.7  6.9  7.1  7.1  7.2  7.3  7.4  7.7   6.1

Arikar   11   25   47   50   51   52   54   55   60   65   73   76   87  110  120    121  126  151  158  172  197  246  268  338  346  410  494  573  580  650   192
         4.9  5.4  5.7  5.7  5.7  5.7  5.7  5.8  5.8  5.9  6.0  6.0  6.1  6.3  6.3    6.3  6.3  6.5  6.5  6.5  6.6  6.7  6.9  7.0  7.2  7.3  7.4  7.7  8.1  8.4   6.5

Tolai     7   18   21   32   59   72   74   86   96   97  109  117  122  127  152    157  162  176  189  192  204  223  312  316  339  367  417  466  478  758   198
         4.6  5.2  5.3  5.5  5.8  5.9  6.0  6.1  6.2  6.2  6.3  6.3  6.3  6.3  6.5    6.5  6.5  6.6  6.6  6.6  6.7  6.8  7.1  7.1  7.2  7.3  7.5  7.6  7.7  9.2   6.5

Guam     16   29   35   37   45   49   62   71   99  108  114  129  141  148  165    166  179  202  231  235  252  261  270  352  390  449  453  490  507  646   211
         5.1  5.5  5.5  5.6  5.7  5.7  5.9  5.9  6.2  6.3  6.3  6.3  6.4  6.4  6.5    6.5  6.6  6.7  6.8  6.8  6.9  6.9  7.0  7.3  7.4  7.6  7.6  7.7  7.8  8.4   6.6

S.Cruz   19   31   34   39   41   56   66   67   79   84   89  113  137  139  150    190  232  249  254  278  298  303  326  337  351  378  435  516  585  618   216
         5.2  5.5  5.5  5.6  5.6  5.8  5.9  5.9  6.0  6.1  6.1  6.3  6.4  6.4  6.4    6.6  6.8  6.9  6.9  7.0  7.1  7.1  7.2  7.2  7.3  7.3  7.5  7.9  8.2  8.3   6.6

Teita     6   17   24   43  100  123  134  144  171  173  174  175  182  191  212    217  221  224  226  264  292  301  318  323  425  439  469  506  584  615   243
         4.6  5.2  5.4  5.7  6.2  6.3  6.4  6.4  6.5  6.6  6.6  6.6  6.6  6.6  6.7    6.7  6.8  6.8  6.8  6.9  7.1  7.1  7.1  7.1  7.5  7.5  7.6  7.8  8.2  8.3   6.7

Peru     20   22   42   44  103  106  116  156  167  193  196  203  208  213  214    229  253  257  263  271  297  324  336  407  419  426  491  513  542  598   254
         5.2  5.3  5.6  5.7  6.2  6.3  6.5  6.5  6.6  6.6  6.7  6.7  6.7  6.7  6.7    6.8  6.9  6.9  7.0  7.1  7.2  7.2  7.4  7.5  7.5  7.7  7.8  8.0  8.2   6.8

Philip   23   33   77   80   83  101  128  131  136  140  146  178  215  245  260    266  281  288  295  334  344  345  356  382  393  440  447  462  509  718   261
         5.3  5.5  6.0  6.0  6.1  6.2  6.3  6.4  6.4  6.4  6.4  6.6  6.7  6.9  6.9    7.0  7.0  7.0  7.1  7.2  7.2  7.3  7.3  7.3  7.4  7.6  7.6  7.6  7.8  8.9   6.8

Zulu      5   46   94  104  112  161  184  201  206  220  236  275  279  290  331    353  359  366  371  427  432  455  508  525  527  544  653  748  800  821   354
         4.5  5.7  6.2  6.3  6.3  6.5  6.6  6.7  6.7  6.8  6.8  7.0  7.0  7.1  7.2    7.3  7.3  7.3  7.3  7.5  7.5  7.6  7.8  7.9  7.9  8.0  8.5  9.1  9.8   10    7.3

Zalava   40   88   91   93  105  132  145  160  168  170  186  242  289  314  363    372  395  396  457  467  502  517  529  530  563  595  609  640  744  779   356
         5.6  6.1  6.1  6.1  6.3  6.4  6.4  6.5  6.5  6.5  6.6  6.9  7.1  7.1  7.3    7.3  7.4  7.4  7.6  7.6  7.8  7.9  7.9  7.9  8.1  8.2  8.3  8.4  9.1  9.5   7.3

Hainan   78   82  125  177  181  185  200  211  307  343  348  349  358  374  377    398  405  412  421  436  456  473  499  536  588  614  617  632  693  695   386
         6.0  6.1  6.3  6.6  6.6  6.6  6.6  6.7  6.7  7.1  7.2  7.3  7.3  7.3  7.3    7.4  7.4  7.4  7.5  7.5  7.6  7.7  7.8  7.9  8.2  8.3  8.3  8.4  8.7  8.7   7.4

Tasman   28   63   68   81  135  195  205  227  233  234  247  258  327  328  354    365  406  459  475  546  551  569  576  579  582  637  676  692  704  816   387
         5.5  5.9  5.9  6.0  6.4  6.6  6.7  6.8  6.8  6.8  6.9  6.9  7.2  7.2  7.3    7.3  7.4  7.6  7.7  8.0  8.0  8.1  8.1  8.1  8.1  8.4  8.6  8.7  8.8   10    7.4

Andama  111  115  118  119  262  276  277  280  284  319  330  381  404  409  413    442  443  454  458  465  484  514  535  537  552  625  672  696  759  764   417
         6.3  6.3  6.3  6.3  6.9  7.0  7.0  7.0  7.0  7.1  7.2  7.3  7.4  7.4  7.4    7.6  7.6  7.6  7.6  7.6  7.7  7.8  7.9  7.9  8.0  8.4  8.6  8.7  9.2  9.3   7.5

Morior    1   36  102  198  225  239  241  282  304  320  355  414  416  424  430    441  444  476  477  486  501  571  587  590  600  603  633  705  735  767   420
         4.0  5.6  6.2  6.7  6.8  6.9  6.9  7.0  7.1  7.1  7.3  7.4  7.4  7.5  7.5    7.6  7.6  7.7  7.7  7.7  7.8  8.1  8.2  8.2  8.2  8.3  8.4  8.8  9.0  9.3   7.5

Norse     9   64   75  187  194  228  259  272  283  299  311  341  383  400  420    438  479  504  510  557  561  572  583  596  604  642  708  724  732  741   423
         4.8  5.9  6.0  6.6  6.6  6.8  6.9  7.0  7.0  7.1  7.1  7.2  7.3  7.4  7.5    7.5  7.7  7.8  7.8  8.0  8.1  8.1  8.2  8.2  8.3  8.4  8.8  8.9  9.0  9.0   7.5

Egypt    53  142  183  188  230  250  256  265  267  313  350  364  384  387  411    445  463  487  512  570  591  592  593  611  638  658  660  691  722  751   434
         5.7  6.4  6.6  6.6  6.8  6.9  6.9  6.9  7.0  7.1  7.3  7.3  7.3  7.3  7.4    7.6  7.6  7.7  7.8  8.1  8.2  8.2  8.2  8.3  8.4  8.5  8.5  8.7  8.9  9.1   7.6

Atayal   70  163  164  169  216  219  237  255  269  294  305  308  369  392  446    519  528  554  575  589  597  599  602  629  631  675  730  757  768    .    435
         5.9  6.5  6.5  6.5  6.7  6.8  6.9  7.0  7.1  7.1  7.1  7.3  7.4  7.6  7.9    7.9  7.9  8.0  8.1  8.2  8.2  8.2  8.4  8.4  8.6  9.0  9.2  9.3    .    7.6

Ainu     57  147  155  159  209  244  291  296  329  335  375  380  385  386  401    422  433  460  533  539  555  607  636  639  649  677  690  728  749  815   443
         5.8  6.4  6.5  6.5  6.7  6.9  7.1  7.1  7.2  7.2  7.3  7.3  7.3  7.3  7.4    7.5  7.5  7.6  7.9  7.9  8.0  8.3  8.4  8.4  8.4  8.6  8.7  8.9  9.1   10    7.7

Anyang    8   13  218  243  251  273  287  325  332  347  368  370  389  391  482    515  547  558  568  616  628  644  647  664  678  687  698  721  727  781   466
         4.7  5.0  6.8  6.9  6.9  7.0  7.0  7.2  7.2  7.3  7.3  7.3  7.4  7.4  7.7    7.8  8.0  8.0  8.1  8.3  8.4  8.4  8.4  8.5  8.6  8.7  8.7  8.9  8.9  9.5   7.7

NJapan    2  153  210  238  248  322  408  423  428  448  451  493  505  532  548    550  560  562  586  606  608  635  651  655  697  736  771  782  791  808   514
         4.0  6.5  6.7  6.9  6.9  7.1  7.4  7.5  7.5  7.6  7.6  7.7  7.8  7.9  8.0    8.0  8.1  8.1  8.2  8.3  8.3  8.4  8.4  8.5  8.7  9.0  9.4  9.6  9.7   10    7.9

SJapan   92  107  285  286  293  309  394  415  429  431  468  489  496  518  521    556  559  605  613  622  674  679  683  703  713  726  772  787  790  802   527
         6.1  6.3  7.0  7.0  7.1  7.1  7.4  7.4  7.5  7.5  7.6  7.7  7.7  7.9  7.9    8.0  8.1  8.3  8.3  8.3  8.6  8.6  8.6  8.8  8.8  8.9  9.4  9.7  9.9  9.9   8.0

Dogon   149  321  361  362  376  379  388  450  452  481  483  511  522  526  531    553  567  574  577  620  621  627  645  684  707  733  753  769  785  812   544
         6.4  7.1  7.3  7.3  7.3  7.3  7.6  7.6  7.6  7.7  7.7  7.8  7.9  7.9  7.9    8.0  8.1  8.1  8.3  8.3  8.4  8.4  8.7  8.8  9.0  9.2  9.4  9.6   10    8.1

Berg     58  124  357  402  403  434  437  461  470  488  538  541  581  594  623    630  659  669  671  688  694  706  712  763  783  788  799  817  818  820   584
         5.8  6.3  7.3  7.4  7.4  7.5  7.5  7.6  7.6  7.7  7.9  8.0  8.1  8.2  8.3    8.4  8.5  8.5  8.6  8.7  8.7  8.8  8.9  9.3  9.6  9.7  9.8   10   10   10    8.4

Hawaii  274  342  471  492  498  523  524  534  543  545  549  566  634  643  654    656  665  667  668  681  720  740  746  754  760  761  776  797  807  809   627
         7.0  7.2  7.6  7.7  7.7  7.9  7.9  7.9  7.9  8.0  8.0  8.0  8.1  8.4  8.4    8.5  8.5  8.5  8.5  8.5  8.6  8.9  9.0  9.1  9.2  9.2  9.3  9.4  9.8   10    8.5

Easter  207  222  306  310  340  397  480  612  626  641  657  661  673  689  701    709  711  725  731  734  743  747  773  792  793  796  803  806  813  814   634
         6.7  6.8  7.1  7.1  7.2  7.4  7.7  8.3  8.4  8.4  8.5  8.5  8.6  8.7  8.8    8.8  8.8  8.9  9.0  9.0  9.0  9.1  9.4  9.7  9.7  9.7  9.9  9.9   10   10    8.6

Bushma  154  373  418  485  500  503  540  565  601  624  662  682  700  702  710    717  729  738  750  752  756  765  766  774  775  777  778  804  828  833   659
         6.5  7.3  7.5  7.7  7.8  7.8  8.0  8.1  8.2  8.4  8.5  8.6  8.7  8.8  8.8    8.9  8.9  9.0  9.1  9.2  9.2  9.3  9.3  9.4  9.4  9.4  9.5  9.9   11   11    8.8

Eskimo  315  464  472  474  495  564  578  610  619  648  652  663  670  686  699    716  719  723  745  755  762  786  789  795  801  810  811  819  830  832   677
         7.1  7.6  7.6  7.7  7.7  8.1  8.1  8.3  8.3  8.4  8.5  8.5  8.6  8.7  8.7    8.9  8.9  8.9  9.1  9.2  9.3  9.6  9.7  9.7  9.8   10   10   10   11   11    8.9

Buriat  302  520  666  680  685  714  715  737  739  742  770  780  784  794  798    805  822  823  824  825  826  827  829  831  834  835  836  837  838  839   762
         7.1  7.9  8.5  8.6  8.7  8.8  8.9  9.0  9.0  9.0  9.4  9.5  9.6  9.7  9.8    9.9   10   10   11   11   11   11   11   11   11   11   11   11   12   12    9.9
```

POPKIN results: Elmenteita B #1922

```
Guam    5   8  10  18  20  21  28  47  49  53  58  59  86  91  92    99 107 135 143 149 150 151 163 184 235 299 309 313 512 618   140
        4.1 4.5 4.5 4.7 4.7 4.7 4.9 5.2 5.2 5.2 5.3 5.3 5.5 5.6 5.6   5.6 5.6 5.8 5.9 5.9 5.9 5.9 6.0 6.2 6.5 6.7 6.7 6.7 7.5 8.1   5.7

Peru    3   4   6  14  19  50  52  67  78  85 105 113 126 128 136   167 197 206 212 214 217 220 267 282 341 345 373 388 404 432   175
        4.1 4.1 4.3 4.6 4.7 5.2 5.2 5.4 5.5 5.5 5.6 5.7 5.8 5.8 5.8   6.1 6.2 6.3 6.3 6.3 6.4 6.4 6.6 6.6 6.8 6.8 7.0 7.0 7.1 7.2   5.9

Egypt   1  11  12  29  30  31  38  41  87  89 137 146 152 190 208   210 223 262 293 314 321 352 374 399 434 447 485 514 541 598   232
        3.8 4.6 4.6 4.9 4.9 4.9 5.1 5.1 5.5 5.5 5.8 5.9 6.0 6.2 6.3   6.3 6.4 6.5 6.6 6.7 6.7 6.9 7.0 7.1 7.2 7.3 7.4 7.5 7.6 8.0   6.1

Morior 13  26  27  39  56  65  76  77  88 148 170 183 231 240 252   268 273 289 294 298 331 357 364 367 375 417 481 602 645 650   257
        4.6 4.8 4.8 5.1 5.2 5.3 5.4 5.4 5.5 5.9 6.1 6.2 6.4 6.5 6.5   6.6 6.6 6.6 6.6 6.6 6.6 6.7 6.8 6.9 6.9 6.9 7.0 7.1 7.4 8.0 8.2 8.3   6.4

Arikar 48  73  74  80  94 101 110 118 124 161 182 196 224 229 232   238 263 278 283 303 325 362 384 387 396 419 457 478 549 625   260
        5.2 5.4 5.4 5.5 5.6 5.6 5.7 5.7 5.7 6.0 6.2 6.2 6.4 6.4 6.4   6.5 6.5 6.6 6.6 6.6 6.7 6.7 6.9 7.0 7.0 7.1 7.1 7.3 7.4 7.7 8.1   6.4

Hawaii  2   7   9  17  23  40  51  62 103 106 117 119 129 242 254   258 259 264 265 296 368 393 406 466 489 566 584 585 632 701   260
        3.9 4.5 4.5 4.7 4.8 5.1 5.2 5.3 5.6 5.6 5.7 5.7 5.8 6.5 6.5   6.5 6.5 6.5 6.5 6.6 6.9 7.1 7.1 7.4 7.4 7.7 7.9 7.9 8.2 8.6   6.3

SJapan 24  66  69  72  98 133 154 157 160 168 172 176 194 199 230   244 280 297 302 308 312 323 445 474 479 503 525 544 574 637   277
        4.8 5.4 5.4 5.4 5.6 5.8 6.0 6.0 6.1 6.1 6.1 6.2 6.2 6.4   4.5 5.4 5.4 5.4 5.6 5.8 6.0 6.0 6.1 6.1 6.1 6.2 6.2 6.4   6.5 6.6 6.7 6.7 6.7 7.3 7.4 7.4 7.5 7.5 7.6 7.8 8.2   6.5

Norse  15  25  32  34  55  68  84 112 114 121 130 139 169 205 255   333 337 350 363 369 380 386 403 486 494 545 546 563 630 709   282
        4.6 4.8 4.9 4.9 5.2 5.4 5.5 5.7 5.7 5.7 5.8 5.8 6.1 6.3 6.5   6.8 6.8 6.9 6.9 6.9 7.0 7.0 7.1 7.4 7.5 7.6 7.6 7.8 8.1 8.6   6.4

Hainan 33  46  61  90 108 111 140 147 164 187 198 204 216 221 227   261 270 301 311 339 348 420 438 472 476 499 543 597 612 661   290
        4.9 5.2 5.3 5.6 5.7 5.7 5.8 5.9 6.1 6.2 6.2 6.3 6.4 6.4 6.4   6.5 6.6 6.7 6.7 6.8 6.8 7.1 7.3 7.4 7.4 7.5 7.6 8.0 8.0 8.3   6.6

Zalava 36  37  43  63  79  93 127 131 159 178 279 281 304 316 320   328 335 336 347 356 389 411 415 422 444 448 470 532 611 729   301
        5.0 5.1 5.1 5.3 5.5 5.6 5.8 5.8 6.0 6.1 6.6 6.6 6.7 6.7 6.7   6.8 6.8 6.8 6.8 6.9 7.0 7.1 7.1 7.2 7.3 7.3 7.4 7.6 8.0 8.9   6.6

S.Cruz 45  57  60  70  71 115 116 166 185 241 245 248 260 269 305   306 322 358 366 378 402 413 418 460 473 495 501 547 551 642   303
        5.2 5.3 5.3 5.4 5.4 5.7 5.7 6.1 6.2 6.5 6.5 6.5 6.6 6.6 6.7   6.7 6.7 6.9 6.9 7.0 7.1 7.1 7.1 7.3 7.4 7.5 7.5 7.6 7.7 8.2   6.6

Philip 16  35 120 132 180 188 191 195 200 209 215 233 234 247 291   334 338 370 372 395 412 453 463 465 520 557 560 579 654 699   329
        4.7 5.0 5.7 5.8 6.2 6.2 6.2 6.2 6.2 6.3 6.4 6.5 6.5 6.5 6.6   6.8 6.8 6.9 7.0 7.1 7.1 7.3 7.3 7.4 7.5 7.7 7.7 7.8 8.3 8.5   6.7

NJapan 54  75  81  82  95 102 104 123 144 173 225 250 253 295 310   317 329 379 401 423 490 500 529 550 595 636 643 673 677 691   340
        5.2 5.4 5.5 5.5 5.6 5.6 5.6 5.7 5.9 6.1 6.4 6.5 6.5 6.6 6.7   6.7 6.8 7.0 7.1 7.2 7.4 7.5 7.6 7.7 8.0 8.2 8.2 8.4 8.4 8.5   6.8

Andama 97 109 130 142 145 179 203 213 237 243 318 319 355 377 383   397 400 439 449 461 467 482 511 517 535 537 592 615 676 731   376
        5.6 5.7 5.8 5.9 5.9 6.2 6.2 6.3 6.5 6.5 6.7 6.7 6.9 7.0 7.0   7.1 7.1 7.3 7.3 7.3 7.4 7.4 7.5 7.5 7.6 7.6 7.9 8.1 8.4 8.9   7.0

Anyang 22  44 141 153 165 171 174 181 226 236 266 286 292 349 353   354 371 390 506 507 509 526 542 554 567 619 633 655 711 766   376
        4.8 5.1 5.9 6.0 6.1 6.1 6.1 6.2 6.4 6.5 6.6 6.6 6.6 6.9 6.9   6.9 6.9 7.1 7.5 7.5 7.5 7.5 7.6 7.7 7.7 7.8 8.1 8.2 8.3 8.7 9.4   7.0

Ainu   64  96 122 175 177 189 256 276 300 307 409 428 443 451 464   475 513 530 538 552 553 565 577 586 589 593 604 664 686 733   431
        5.3 5.6 5.7 6.1 6.1 6.2 6.6 6.6 6.7 6.7 7.1 7.2 7.3 7.3 7.3   7.4 7.5 7.6 7.6 7.7 7.7 7.7 7.8 7.9 7.9 7.9 8.0 8.3 8.5 8.9   7.2

Teita  42  83 158 186 192 324 382 407 414 436 437 442 459 471 504   521 527 569 573 583 587 594 600 603 609 651 663 672 716 737   471
        5.1 5.5 6.0 6.2 6.2 6.7 7.0 7.1 7.1 7.2 7.3 7.3 7.3 7.4 7.5   7.5 7.5 7.7 7.8 7.8 7.9 8.0 8.0 8.0 8.0 8.3 8.3 8.4 8.7 8.9   7.4

Atayal 125 193 219 222 246 340 344 351 425 433 446 468 477 492 496   502 515 531 562 581 588 590 610 622 667 688 726 741 762   .   481
        5.8 6.2 6.4 6.4 6.5 6.8 6.8 6.9 7.2 7.2 7.3 7.4 7.4 7.5 7.5   7.5 7.5 7.6 7.7 7.8 7.9 7.9 8.0 8.1 8.4 8.5 8.8 9.0 9.4   .   7.5

Berg   155 156 207 211 239 272 274 288 327 342 416 455 498 505 510   522 540 576 601 613 616 639 640 658 697 725 752 760 769 784   492
        6.0 6.0 6.3 6.3 6.5 6.6 6.6 6.6 6.8 6.8 7.1 7.3 7.5 7.5 7.5   7.5 7.6 7.8 8.0 8.0 8.1 8.2 8.2 8.3 8.5 8.8 9.2 9.3 9.5 9.7   7.6

Eskimo 134 201 249 257 275 285 326 361 376 392 405 429 430 431 452   484 523 539 629 652 679 680 682 707 713 721 724 744 786 794   502
        5.8 6.2 6.5 6.5 6.6 6.6 6.8 6.9 7.0 7.1 7.1 7.2 7.2 7.2 7.3   7.4 7.5 7.6 8.1 8.3 8.4 8.4 8.5 8.6 8.7 8.8 8.8 8.9 9.1 9.8 9.9   7.7

Zulu   271 287 290 315 365 381 391 398 410 440 450 483 491 508 516   534 555 570 572 582 626 649 657 665 670 704 712 718 719 811   525
        6.6 6.6 6.6 6.7 6.9 7.0 7.1 7.1 7.1 7.3 7.3 7.4 7.4 7.5 7.5   7.6 7.7 7.7 7.7 7.8 8.1 8.2 8.3 8.3 8.4 8.6 8.7 8.7 8.8   10   7.7

Easter 100 228 251 346 360 385 426 456 458 469 480 487 497 518 519   564 568 578 591 605 606 621 634 641 690 728 743 745 750 771   527
        5.6 6.4 6.5 6.8 6.9 7.0 7.2 7.3 7.3 7.4 7.4 7.4 7.5 7.5 7.5   7.7 7.7 7.7 7.8 7.9 8.0 8.0 8.1 8.2 8.2 8.5 8.8 9.1 9.1 9.2 9.5   7.7

Tolai  202 330 343 408 424 462 493 524 528 548 559 575 580 614 623   624 647 656 671 674 681 684 703 705 714 739 749 757 767 775   592
        6.2 6.8 6.8 7.1 7.2 7.3 7.5 7.5 7.5 7.6 7.7 7.8 7.8 8.0 8.1   8.1 8.2 8.3 8.4 8.4 8.5 8.5 8.6 8.6 8.7 9.0 9.2 9.3 9.4 9.5   8.1

Dogon  284 394 427 435 441 454 533 556 561 571 596 644 648 662 678   683 700 723 732 736 742 754 755 756 759 770 773 787 799 800   638
        6.6 7.1 7.2 7.2 7.3 7.3 7.6 7.7 7.7 7.7 7.8 8.0 8.2 8.2 8.3 8.4   8.5 8.6 8.8 8.9 8.9 9.0 9.2 9.3 9.3 9.3 9.5 9.5 9.8 9.9   10   8.4

Buriat 218 277 421 488 558 635 638 646 660 668 675 685 695 698 706   710 722 735 740 748 758 763 774 776 782 790 812 818 829 834   675
        6.4 6.6 7.1 7.4 7.7 8.2 8.2 8.2 8.3 8.4 8.4 8.5 8.5 8.5 8.6   8.7 8.8 8.9 9.0 9.1 9.3 9.4 9.5 9.6 9.7 9.8  10  11  11  11   8.8

Austra 536 607 627 628 631 653 689 692 696 708 715 717 720 727 730   746 751 753 761 779 780 783 793 795 796 804 817 825 831 835   731
        7.6 8.0 8.1 8.1 8.2 8.3 8.5 8.5 8.5 8.6 8.7 8.7 8.8 8.8 8.9   9.1 9.2 9.3 9.6 9.6 9.7 9.9 9.9 9.9 10  11  11  11  11   9.2

Tasman 162 332 359 608 617 693 694 747 765 768 778 781 788 801 803   805 809 810 813 814 819 821 822 823 824 830 833 836 837 839   734
        6.0 6.8 6.9 8.0 8.1 8.5 8.5 9.1 9.4 9.4 9.6 9.7 9.8  10  10   10  10  10  10  10  11  11  11  11  11  11  11  12  12  13   9.8

Bushma 599 620 659 666 669 687 702 734 738 764 772 777 785 789 791   792 797 798 802 806 807 808 815 816 820 826 827 828 832 838   765
        8.0 8.1 8.3 8.4 8.4 8.5 8.6 8.9 9.0 9.4 9.5 9.6 9.8 9.8 9.8   9.8 9.9 9.9 10  10  10  10  10  11  11  11  11  11  12   9.7
```

TABLE 9C-7
POPKIN results: Nakuru 9 #1923

```
Austra    1    5    6    8   10   15   18   24   31   32   40   42   49   68   84    89   90   93  103  114  155  166  207  220  246  269  347  364  393  421   124
        5.1  5.5  5.5  5.6  5.9  6.0  6.1  6.2  6.4  6.4  6.5  6.5  6.6  6.8  6.9   6.9  6.9  7.0  7.0  7.1  7.3  7.4  7.6  7.7  7.8  7.9  8.2  8.3  8.5  8.6   6.9

Egypt     9   17   19   27   28   33   35   54   56   61   62   74   86   97   99   115  117  119  125  134  141  143  151  162  221  223  249  294  357  453   125
        5.6  6.0  6.1  6.3  6.3  6.4  6.4  6.6  6.6  6.7  6.7  6.8  6.9  7.0  7.0   7.1  7.1  7.1  7.2  7.2  7.2  7.2  7.3  7.4  7.7  7.7  7.8  8.0  8.3  8.8   7.0

Peru      3   12   29   34   37   47   51   53   55   66   81   83   85   88   91    94  153  157  191  202  206  208  210  268  287  299  332  339  348  381   150
        5.5  5.9  6.3  6.4  6.4  6.5  6.6  6.6  6.6  6.7  6.9  6.9  6.9  6.9  6.9   7.0  7.3  7.3  7.5  7.6  7.6  7.6  7.6  7.9  8.0  8.0  8.2  8.2  8.3  8.4   7.2

S.Cruz    4   11   23   44   45   50   58   59   77   96  122  131  137  148  150   160  175  176  199  203  228  264  266  282  285  298  312  353  435  465   175
        5.5  5.9  6.2  6.5  6.5  6.6  6.6  6.7  6.8  7.0  7.1  7.2  7.2  7.3  7.3   7.3  7.5  7.5  7.6  7.6  7.7  7.9  7.9  8.0  8.0  8.0  8.1  8.3  8.7  8.8   7.3

Norse    14   36   39   41   46   52   75  100  109  121  126  128  133  135  154   170  183  196  224  232  234  240  252  267  308  368  390  402  409  475   189
        5.9  6.4  6.5  6.5  6.5  6.6  6.8  7.0  7.0  7.1  7.2  7.2  7.2  7.2  7.3   7.4  7.5  7.5  7.7  7.7  7.7  7.8  7.8  7.9  8.1  8.3  8.4  8.5  8.5  8.9   7.4

Tolai     2   38   57   63   67   73   79  107  139  144  146  156  161  164  201   217  222  261  271  281  286  323  350  363  374  403  406  428  589  591   232
        5.3  6.5  6.6  6.7  6.7  6.8  6.9  7.0  7.2  7.2  7.3  7.3  7.3  7.4  7.6   7.6  7.7  7.9  7.9  8.0  8.0  8.1  8.3  8.3  8.4  8.5  8.5  8.6  9.4  9.4   7.6

Zalava   13   22   43   64   71   72   95  120  124  142  163  172  173  174  193   194  233  241  270  280  296  306  338  386  395  411  415  527  533  535   233
        5.9  6.2  6.5  6.7  6.8  6.8  7.0  7.1  7.2  7.2  7.4  7.4  7.4  7.4  7.5   7.5  7.7  7.8  7.9  8.0  8.0  8.1  8.2  8.4  8.5  8.5  8.6  9.1  9.2  9.2   7.6

Teita     7  105  112  116  123  130  147  177  180  182  214  235  238  245  250   256  273  291  297  301  333  342  345  394  405  448  456  500  570  671   278
        5.6  7.0  7.1  7.1  7.1  7.2  7.3  7.5  7.5  7.5  7.6  7.7  7.7  7.8  7.8   7.8  7.9  8.0  8.0  8.1  8.2  8.2  8.2  8.5  8.5  8.7  8.8  9.0  9.3  9.9   7.9

Zulu     25   30   69   80   92  104  113  127  132  200  212  216  239  251  278   311  355  360  365  379  384  408  464  476  477  489  505  520  620  696   299
        6.2  6.4  6.8  6.9  6.9  7.0  7.1  7.2  7.2  7.6  7.6  7.6  7.7  7.8  7.9   8.1  8.3  8.3  8.3  8.4  8.4  8.5  8.8  8.9  8.9  9.0  9.0  9.1  9.6   10   8.0

Ainu     21   78   98  152  158  159  169  186  187  195  205  242  248  259  272   276  290  295  307  331  372  385  413  458  466  509  529  603  687  706   309
        6.1  6.8  7.0  7.3  7.3  7.3  7.4  7.5  7.5  7.5  7.6  7.8  7.8  7.8  7.9   7.9  8.0  8.0  8.1  8.2  8.4  8.4  8.5  8.8  8.8  8.9  9.0  9.1  9.5   10   10   8.1

Dogon    20   60   87  145  189  213  218  226  229  255  260  263  284  305  319   346  376  378  388  389  429  459  479  506  522  541  585  593  693  712   349
        6.1  6.7  6.9  7.2  7.5  7.6  7.7  7.7  7.7  7.8  7.9  7.9  8.0  8.1  8.1   8.2  8.4  8.4  8.4  8.4  8.6  8.8  8.9  9.0  9.1  9.2  9.4  9.4   10   10   8.3

Andama   76  111  165  168  171  178  185  190  225  254  279  321  327  366  367   383  416  438  442  481  487  501  503  537  555  558  624  655  658  762   379
        6.8  7.1  7.4  7.4  7.4  7.5  7.5  7.5  7.7  7.8  7.9  8.1  8.2  8.3  8.3   8.4  8.6  8.7  8.7  8.9  8.9  9.0  9.0  9.2  9.3  9.3  9.6  9.8  9.8   11   8.4

Atayal   16   48  129  209  231  236  265  274  309  313  328  370  392  404  420   439  452  460  493  497  551  567  576  600  610  629  650  678  690    .   408
        6.0  6.6  7.2  7.6  7.7  7.7  7.9  7.9  8.1  8.1  8.2  8.4  8.5  8.5  8.6   8.7  8.8  8.8  9.0  9.0  9.3  9.3  9.4  9.5  9.5  9.6  9.7   10   10    .   8.5

Arikar  106  108  192  198  275  288  302  303  329  334  337  356  398  399  400   419  427  446  447  451  549  563  606  616  617  619  632  662  680  721   426
        7.0  7.0  7.5  7.6  7.9  8.0  8.1  8.1  8.2  8.2  8.2  8.3  8.5  8.5  8.5   8.6  8.6  8.7  8.7  8.9  9.2  9.3  9.5  9.5  9.5  9.6  9.6  9.8   10   10   8.6

Guam     70  140  167  179  244  310  320  322  324  330  336  412  417  426  440   473  484  491  498  518  534  542  543  548  564  584  614  643  783  797   435
        6.8  7.2  7.4  7.5  7.8  8.1  8.1  8.1  8.1  8.2  8.2  8.5  8.6  8.6  8.7   8.8  8.9  9.0  9.0  9.1  9.2  9.2  9.2  9.3  9.4  9.5  9.7  9.9   11   11   8.7

Hainan  204  230  243  253  257  258  292  293  300  318  325  343  361  373  397   449  463  467  471  474  530  562  581  587  608  672  675  705  736  742   439
        7.6  7.7  7.8  7.8  7.8  7.8  8.0  8.0  8.0  8.1  8.1  8.2  8.3  8.4  8.5   8.7  8.8  8.9  8.9  8.9  9.1  9.3  9.4  9.4  9.5  9.9   10   10   11   11   8.7

NJapan   26   65  237  262  314  315  340  358  369  377  424  433  437  455  478   492  512  515  524  538  559  574  601  618  621  634  638  639  649  785   460
        6.2  6.7  7.7  7.9  8.1  8.1  8.2  8.3  8.3  8.4  8.6  8.7  8.7  8.8  8.9   9.0  9.0  9.1  9.1  9.2  9.3  9.3  9.5  9.5  9.6  9.7  9.7  9.7  9.7   11   8.8

Berg     82  101  184  215  219  247  335  371  401  414  418  430  443  482  486   511  519  550  561  565  592  628  652  679  684  725  727  761  766  782   484
        6.9  7.0  7.5  7.6  7.7  7.8  8.2  8.4  8.5  8.6  8.6  8.6  8.7  8.9  8.9   9.0  9.1  9.2  9.3  9.3  9.4  9.6  9.8   10   10   10   10   11   11   11   9.0

Philip  188  277  317  341  351  375  387  422  436  441  470  472  480  490  499   508  513  546  547  553  560  566  568  583  605  630  645  669  708  792   498
        7.5  7.9  8.1  8.2  8.3  8.4  8.4  8.6  8.7  8.7  8.9  8.9  8.9  9.0  9.0   9.0  9.1  9.2  9.2  9.3  9.3  9.3  9.3  9.4  9.5  9.6  9.7  9.9   10   11   9.0

SJapan  118  211  283  304  316  349  362  423  425  431  432  462  494  517  521   540  545  573  582  590  637  648  663  664  668  709  726  729  739  777   515
        7.1  7.6  8.0  8.1  8.1  8.3  8.3  8.6  8.6  8.7  8.7  8.8  9.0  9.1  9.1   9.2  9.2  9.3  9.4  9.4  9.7  9.7  9.9  9.9  9.9   10   10   10   11   11   9.1

Morior  102  138  326  344  380  391  468  488  496  516  523  526  531  532  554   556  569  580  609  611  627  646  681  692  698  715  734  760  770  790   545
        7.0  7.2  8.2  8.2  8.4  8.4  8.9  9.0  9.0  9.1  9.1  9.1  9.2  9.2  9.3   9.3  9.3  9.4  9.5  9.5  9.6  9.7   10   10   10   11   11   11   11   11   9.3

Tasman  110  136  149  354  359  396  457  502  504  510  514  595  612  615  622   636  640  657  665  670  677  694  699  702  716  720  724  774  795  819   567
        7.0  7.2  7.3  8.3  8.3  8.5  8.8  9.0  9.0  9.0  9.1  9.4  9.5  9.5  9.6   9.7  9.7  9.8  9.9  9.9   10   10   10   10   10   10   10   11   11   12   9.5

Eskimo  181  289  352  407  461  483  485  495  539  557  572  575  577  588  594   602  641  644  647  659  674  682  686  688  710  737  738  751  755  799   586
        7.5  8.0  8.3  8.5  8.8  8.9  8.9  9.0  9.2  9.3  9.3  9.4  9.4  9.4  9.4   9.5  9.7  9.7  9.7  9.8  9.9   10   10   10   10   11   11   11   11   11   9.5

Anyang  197  227  410  525  528  544  552  578  579  597  599  604  625  626  631   660  676  685  689  695  697  714  718  722  728  768  771  776  780  800   623
        7.6  7.7  8.5  9.1  9.1  9.2  9.3  9.4  9.4  9.4  9.5  9.5  9.6  9.6  9.6   9.8   10   10   10   10   10   10   10   10   10   11   11   11   11   11   9.8

Bushma  382  445  450  507  598  607  623  642  653  656  661  667  691  703  719   723  740  745  750  752  753  764  772  779  788  796  802  803  807  829   687
        8.4  8.7  8.8  9.0  9.5  9.5  9.6  9.7  9.8  9.8  9.8  9.9   10   10   10    10   11   11   11   11   11   11   11   11   11   11   11   11   12   13  10.3

Easter  434  454  536  571  586  596  635  701  711  713  717  730  732  733  735   743  748  757  758  763  778  784  786  789  793  804  805  812  813  826   711
        8.7  8.8  9.2  9.3  9.4  9.4  9.7   10   10   10   10   10   10   11   11    11   11   11   11   11   11   11   11   11   11   12   12   12   12   13  10.5

Hawaii  469  613  633  651  654  666  683  700  704  731  746  747  749  754  756   759  765  767  769  773  775  787  791  798  810  814  818  821  823  828   738
        8.9  9.5  9.6  9.7  9.8  9.9   10   10   10   10   11   11   11   11   11    11   11   11   11   11   11   11   11   11   12   12   12   12   13   13  10.8

Buriat  444  673  707  741  744  781  794  801  806  808  809  811  815  816  817   820  822  824  825  827  830  831  832  833  834  835  836  837  838  839   794
        8.7  9.9   10   11   11   11   11   11   12   12   12   12   12   12   12    12   13   13   13   13   13   13   13   13   13   13   13   14   14   15  12.1
```

TABLE 9C-8

POPKIN results: Willey's Kopje 3 #1925

```
Ainu      8  11  19  34  41  46  51  58  62  71  83  95 106 113 117   165 180 260 276 286 292 300 306 341 346 352 385 400 417 478   190
        4.5 4.8 5.0 5.1 5.2 5.3 5.3 5.4 5.4 5.5 5.6 5.6 5.7 5.7 5.8   6.0 6.0 6.4 6.5 6.5 6.5 6.6 6.6 6.7 6.7 6.7 6.9 6.9 7.0 7.2   6.0

Guam      2   3   9  26  27  32  48  79  84  88  90  93 124 131 170   182 186 192 205 215 221 230 278 365 374 439 553 555 610 661   209
        4.3 4.4 4.7 5.0 5.0 5.1 5.3 5.6 5.6 5.6 5.6 6.0   6.0 6.1 6.1 6.2 6.2 6.2 6.3 6.5 6.8 6.8 7.1 7.6 7.6 7.8 8.1   6.0

Zalava    6  12  24  43  49  54  57  86  96  99 129 135 164 174 185   190 216 239 247 262 293 294 310 324 362 388 407 436 453 588   211
        4.5 4.8 5.0 5.2 5.3 5.3 5.4 5.6 5.7 5.7 5.8 5.9 6.0 6.0 6.1   6.1 6.2 6.3 6.3 6.4 6.5 6.5 6.6 6.6 6.8 6.9 6.9 7.0 7.1 7.7   6.1

Easter    1   4  15  17  18  28  60  81  97 125 126 143 154 162 171   184 193 204 225 227 234 252 283 332 367 476 500 501 536 608   211
        3.8 4.5 4.8 4.9 4.9 5.1 5.4 5.6 5.7 5.8 5.8 5.9 6.0 6.0 6.0   6.1 6.1 6.2 6.2 6.2 6.3 6.4 6.5 6.7 6.8 7.2 7.3 7.3 7.5 7.8   6.0

NJapan   14  30  36  53  63  66  76  82  87 108 116 137 149 210 212   238 248 281 301 330 364 392 405 472 486 494 504 673 682 684   266
        4.8 5.1 5.1 5.3 5.4 5.5 5.5 5.6 5.6 5.7 5.8 5.9 6.2 6.2       6.3 6.3 6.5 6.6 6.6 6.8 6.9 6.9 7.2 7.2 7.3 7.3 8.1 8.3 8.3   6.3

Eskimo   21  39  73 104 111 112 132 133 142 161 167 218 224 226 246   250 291 316 328 344 347 373 380 390 418 489 574 589 639 666   283
        5.0 5.2 5.5 5.7 5.7 5.7 5.9 5.9 5.9 6.0 6.0 6.2 6.2 6.2 6.3   6.3 6.5 6.6 6.6 6.6 6.7 6.7 6.8 6.8 6.9 7.0 7.3 7.7 7.8 8.0 8.1   6.4

SJapan    5  31  40  50  68  80  85 118 134 136 147 187 191 232 251   258 264 265 275 287 314 432 485 547 548 549 590 623 703 748   288
        4.5 5.1 5.2 5.3 5.5 5.6 5.6 5.8 5.9 5.9 5.9 6.1 6.1 6.3 6.3   6.4 6.4 6.4 6.5 6.6 6.7 7.2 7.5 7.5 7.5 7.7 7.7 7.9 8.4 8.8   6.4

Atayal   33  47 110 122 123 128 151 166 169 179 200 242 255 271 337   342 345 348 372 382 403 424 470 471 479 551 559 564 605    .   305
        5.1 5.3 5.7 5.8 5.8 5.8 5.9 6.0 6.0 6.0 6.1 6.3 6.4 6.4 6.7   6.7 6.7 6.7 6.8 6.8 6.9 7.0 7.2 7.2 7.2 7.5 7.6 7.6 7.8    .   6.5

Zulu      7  37  56  91 105 121 153 158 160 172 231 235 277 279 282   284 318 333 363 369 384 414 415 431 435 525 562 620 645 773   308
        4.5 5.1 5.4 5.6 5.7 5.8 5.9 6.0 6.0 6.0 6.3 6.3 6.5 6.5 6.5   6.5 6.6 6.7 6.8 6.8 6.8 6.9 7.0 7.0 7.0 7.4 7.6 7.8 8.0 9.0   6.5

Tolai    10  25  94  98 175 195 198 214 237 241 245 267 297 302 327   391 396 413 440 458 508 510 520 521 529 613 626 647 695 732   364
        4.7 5.0 5.6 5.7 6.0 6.1 6.1 6.2 6.3 6.3 6.3 6.4 6.6 6.6 6.6   6.9 6.9 6.9 7.1 7.1 7.3 7.3 7.4 7.4 7.5 7.8 7.9 8.0 8.3 8.7   6.8

Teita    23  45  52  78 109 197 199 201 219 228 299 311 375 378 386   404 411 421 422 429 437 515 516 546 557 596 600 671 687 745   369
        5.0 5.3 5.3 5.6 5.7 6.1 6.1 6.2 6.2 6.2 6.6 6.6 6.8 6.8 6.9   6.9 6.9 7.0 7.0 7.0 7.0 7.4 7.4 7.5 7.6 7.7 7.8 8.1 8.3 8.8   6.8

Norse    22  61  69  77 130 155 159 178 220 223 240 254 259 336 401   402 427 430 448 461 526 539 643 644 651 658 663 674 675 721   378
        5.0 5.4 5.5 5.6 5.8 6.0 6.0 6.0 6.2 6.2 6.3 6.4 6.4 6.7 6.9   6.9 7.0 7.0 7.1 7.1 7.4 7.5 8.0 8.0 8.0 8.1 8.1 8.1 8.1 8.6   6.8

Anyang   42  72  89 120 150 157 189 208 249 269 273 381 389 397 409   412 449 454 457 462 463 466 498 540 542 598 631 642 709 752   382
        5.2 5.5 5.6 5.8 5.9 6.0 6.1 6.2 6.3 6.4 6.4 6.8 6.9 6.9 6.9   6.9 7.1 7.1 7.1 7.1 7.2 7.2 7.3 7.5 7.5 7.8 7.9 8.0 8.5 8.8   6.9

Philip   16  14  67 114 115 137 160 206 276 305 310 313 380 381 400   405 455 507 517 526 566 567 575 578 604 616 631 660 677 686   390
        4.8 5.2 5.5 5.7 5.8 5.9 6.0 6.2 6.5 6.6 6.6 6.6 6.6 6.7 7.0   7.0 7.1 7.3 7.4 7.4 7.5 7.6 7.7 7.7 7.8 7.8 7.9 8.1 8.2 8.3   6.9

Hainan   38  59  64 127 148 207 211 272 308 334 340 343 360 377 408   416 438 452 477 482 561 570 572 583 592 597 637 669 719 761   404
        5.1 5.4 5.4 5.8 5.9 6.2 6.2 6.4 6.6 6.7 6.7 6.7 6.8 6.8 6.9   7.0 7.0 7.1 7.2 7.2 7.6 7.6 7.7 7.7 7.7 7.7 7.9 8.1 8.6 8.9   7.0

Peru     65  74 100 140 156 176 253 257 266 368 370 379 410 442 445   451 469 484 499 505 541 545 556 577 601 607 646 678 689 711   415
        5.4 5.5 5.7 5.9 6.0 6.0 6.4 6.4 6.4 6.8 6.8 6.8 6.9 7.1 7.1   7.1 7.2 7.2 7.3 7.5 7.5 7.5 7.6 7.7 7.8 7.8 8.0 8.2 8.3 8.5   7.0

Hawaii   13  35 101 144 146 196 243 263 319 354 356 383 387 444 446   468 490 506 519 530 535 625 627 628 640 654 679 706 707 710   428
        4.8 5.1 5.7 5.9 5.9 6.1 6.3 6.4 6.6 6.7 6.8 6.8 6.9 7.1 7.1   7.2 7.3 7.3 7.4 7.5 7.5 7.9 7.9 7.9 8.0 8.0 8.2 8.5 8.5 8.5   7.1

Arikar   29  75 163 183 222 256 261 274 315 322 323 353 357 447 480   495 527 532 534 579 580 585 586 591 595 612 632 676 726 759   436
        5.1 5.5 6.0 6.1 6.2 6.4 6.4 6.5 6.6 6.6 6.6 6.7 6.8 7.1 7.2   7.3 7.4 7.5 7.5 7.7 7.7 7.7 7.7 7.7 7.7 7.8 7.9 8.2 8.6 8.9   7.1

Egypt   102 107 194 209 217 304 309 329 331 335 338 351 355 376 423   443 507 514 518 531 582 611 629 641 649 665 668 680 704 714   445
        5.7 5.7 6.1 6.2 6.2 6.6 6.6 6.6 6.7 6.7 6.7 6.7 6.8 6.8 7.0   7.1 7.3 7.4 7.4 7.5 7.7 7.8 7.9 8.0 8.0 8.1 8.1 8.2 8.4 8.5   7.1

Austra  152 181 188 203 213 307 339 349 358 366 395 474 481 488 492   496 497 517 533 560 571 616 633 662 693 697 705 763 770 793   483
        5.9 6.0 6.1 6.2 6.2 6.6 6.7 6.7 6.8 6.8 6.9 7.2 7.2 7.3 7.3   7.3 7.3 7.4 7.5 7.6 7.7 7.8 7.9 8.1 8.3 8.4 8.4 8.9 9.0 9.5   7.4

Morior   20  92 138 229 244 285 289 290 350 361 399 428 475 502 537   554 593 603 606 617 622 630 634 659 718 725 728 729 755 775   486
        5.0 5.6 5.9 6.2 6.3 6.5 6.6 6.7 6.8 6.9 7.2 7.2 7.3 7.5   6.7 7.7 7.8 7.8 7.9 7.9 7.9 8.1 8.6 8.6 8.6 8.8 8.9 9.1 7.4

S.Cruz  141 173 233 268 288 295 317 326 359 393 394 398 450 487 491   523 538 543 558 563 573 614 638 688 708 722 735 737 777 779   490
        5.9 6.0 6.3 6.4 6.5 6.5 6.6 6.6 6.8 6.9 6.9 6.9 7.1 7.3 7.3   7.4 7.5 7.5 7.6 7.6 7.7 7.8 7.9 8.3 8.5 8.6 8.7 8.7 9.1 9.2   7.4

Berg     55  70 177 202 270 280 298 303 325 426 434 459 460 464 473   511 528 565 599 609 652 653 690 701 716 733 744 764 787 797   492
        5.3 5.5 6.0 6.2 6.4 6.5 6.6 6.6 6.6 7.0 7.0 7.1 7.1 7.2 7.2   7.4 7.4 7.6 7.8 7.8 8.0 8.0 8.3 8.4 8.5 8.7 8.8 9.0 9.4 9.6   7.4

Tasman  103 119 145 236 419 509 569 576 619 635 670 672 683 702 712   717 730 731 739 742 753 756 762 768 780 790 796 798 810 834   629
        5.7 5.8 5.9 6.3 7.0 7.3 7.6 7.7 7.8 7.9 8.1 8.1 8.3 8.4 8.5   8.5 8.6 8.7 8.7 8.7 8.8 8.9 8.9 9.0 9.2 9.4 9.5 9.6 9.9  11   8.3

Andama  371 406 483 567 581 648 656 681 691 694 699 713 720 723 734   738 740 743 746 749 767 772 781 783 785 799 807 814 817 824   701
        6.8 6.9 7.2 7.6 7.7 8.0 8.0 8.2 8.3 8.3 8.4 8.5 8.6 8.6 8.7   8.7 8.7 8.8 8.8 8.9 9.0 9.0 9.2 9.3 9.3 9.6 9.8  10  10  10   8.6

Dogon   433 456 467 522 566 587 602 650 657 664 667 685 700 750 751   760 766 789 792 805 811 812 813 815 816 823 825 828 830 835   709
        7.0 7.1 7.2 7.4 7.6 7.7 7.8 8.0 8.1 8.1 8.3 8.4 8.8 8.8       8.9 9.0 9.4 9.5 9.7 9.9 9.9 9.9  10  10  10  10  11  11  11   8.9

Bushma  441 512 568 584 618 624 636 692 698 715 724 727 736 741 758   765 769 774 784 791 795 802 804 806 809 819 820 821 833 838   727
        7.1 7.4 7.6 7.7 7.8 7.9 7.9 8.3 8.4 8.5 8.6 8.6 8.7 8.7 8.9   9.0 9.0 9.0 9.3 9.5 9.5 9.7 9.7 9.8 9.8  10  10  10  11  12   9.0

Buriat  465 493 550 594 655 696 747 754 757 771 776 778 782 786 788   794 800 801 803 808 818 822 826 827 829 831 832 836 837 839   757
        7.2 7.3 7.5 7.7 8.0 8.3 8.8 8.8 8.9 9.0 9.1 9.2 9.3 9.3 9.4   9.5 9.7 9.7 9.7 9.8  10  10  10  11  11  11  11  11  11  12   9.5
```

TABLE 9C-9

POPKIN results: Afalou 5 #2076

```
Philip    1    2    7   11   13   14   16   27   33   34   37   53   59   60   66     70   74   76   80   82  115  134  141  168  175  200  224  272  314  326      96
        3.5  3.8  4.3  4.4  4.5  4.5  4.5  4.9  5.0  5.0  5.0  5.2  5.3  5.3  5.4    5.4  5.5  5.5  5.5  5.6  5.8  6.0  6.0  6.2  6.2  6.3  6.3  6.6  6.7  6.8     5.4

Guam     10   12   25   26   31   41   48   51   54   77   88   91   96  105  121    124  126  127  136  148  267  277  280  285  297  336  414  426  488  516     171
        4.4  4.4  4.9  4.9  4.9  5.1  5.2  5.2  5.3  5.5  5.6  5.6  5.7  5.9    5.9  5.9  5.9  6.0  6.0  6.5  6.6  6.6  6.6  6.7  6.8  7.1  7.1  7.4  7.5     5.9

NJapan   24   43   56   58   72   89   97   99  113  120  123  132  145  181  187    207  223  226  228  238  243  250  290  323  343  363  368  387  435  702     212
        4.8  5.1  5.3  5.3  5.4  5.6  5.6  5.7  5.8  5.8  5.9  5.9  6.0  6.2  6.2    6.3  6.3  6.4  6.4  6.4  6.4  6.4  6.5  6.6  6.8  6.8  6.9  6.9  7.0  7.2  8.3     6.2

Zulu      3    5    8   17   35   46   63   73   85   92   93  111  129  157  193    225  234  257  266  288  300  305  333  381  396  476  528  563  604  627     230
        3.9  4.1  4.4  4.5  5.0  5.1  5.4  5.5  5.6  5.6  5.6  5.8  5.9  6.1  6.3    6.4  6.4  6.5  6.5  6.6  6.6  6.7  6.7  6.8  6.9  7.0  7.3  7.5  7.6  7.8  7.9     6.1

Tolai    19   30   44   62   71  108  109  116  117  135  143  146  152  159  188    204  232  258  261  294  308  328  359  382  409  466  475  487  622  672     245
        4.7  4.9  5.1  5.4  5.4  5.7  5.7  5.8  5.8  6.0  6.0  6.0  6.1  6.1  6.2    6.3  6.4  6.5  6.5  6.7  6.7  6.8  6.9  6.9  7.0  7.3  7.3  7.4  7.8  8.1     6.3

Zalava   42   47   57   79   83   90   95  110  112  118  122  144  158  185  197    222  237  269  316  340  353  357  367  369  424  438  504  557  558  720     256
        5.1  5.2  5.3  5.5  5.6  5.6  5.6  5.7  5.8  5.8  5.9  6.0  6.1  6.2  6.3    6.3  6.4  6.6  6.7  6.8  6.9  6.9  6.9  6.9  7.1  7.2  7.4  7.6  7.6  8.4     6.4

Ainu      9   29   45   52   64   65   68   75   81  151  153  177  191  195  220    239  241  254  264  273  334  344  350  497  502  532  535  653  675  681     265
        4.4  4.9  5.1  5.2  5.4  5.4  5.4  5.5  5.6  6.1  6.1  6.2  6.2  6.3  6.3    6.4  6.4  6.5  6.5  6.6  6.8  6.8  6.9  7.4  7.4  7.5  7.5  8.0  8.1  8.2     6.4

Tasman   20   21   23   28   36   50  103  128  137  173  218  235  293  313  330    341  365  392  425  444  462  500  545  562  566  586  642  647  716  771     336
        4.7  4.8  4.8  4.9  5.0  5.2  5.7  5.9  6.0  6.2  6.3  6.4  6.7  6.7  6.8    6.8  6.9  7.0  7.1  7.2  7.3  7.4  7.6  7.6  7.6  7.7  7.9  8.0  8.4  8.8     6.6

SJapan   15   38   84   86  138  142  147  166  221  309  317  366  402  412  415    416  417  420  432  442  443  445  449  493  499  515  534  596  635  832     361
        4.5  5.0  5.6  5.6  6.0  6.0  6.0  6.1  6.3  6.7  6.7  6.9  7.0  7.1  7.1    7.1  7.1  7.1  7.2  7.2  7.2  7.2  7.2  7.4  7.4  7.5  7.5  7.7  7.9   10     6.9

Hainan   39   40   55  101  162  164  169  198  211  242  253  256  286  358  361    364  370  372  455  463  483  503  507  561  602  612  644  695  705  747     366
        5.1  5.1  5.3  5.7  6.1  6.1  6.2  6.3  6.3  6.4  6.5  6.5  6.6  6.9  6.9    6.9  6.9  6.9  7.2  7.3  7.4  7.4  7.4  7.6  7.8  7.8  8.0  8.2  8.3  8.6     6.9

Egypt    67  119  139  140  154  167  182  262  274  282  291  296  299  304  347    348  352  388  459  464  469  478  479  505  514  548  609  661  699  759     369
        5.4  5.8  6.0  6.0  6.1  6.2  6.2  6.5  6.6  6.6  6.6  6.7  6.7  6.7  6.8    6.8  6.9  7.0  7.2  7.3  7.3  7.3  7.4  7.5  7.6  7.8  8.0  8.3  8.7     6.9

Hawaii   78  155  156  165  174  180  201  227  236  245  247  260  271  275  301    360  384  423  427  446  468  486  520  544  598  625  634  640  659  726     370
        5.5  6.1  6.1  6.1  6.2  6.2  6.3  6.4  6.4  6.4  6.4  6.5  6.6  6.6  6.7    6.9  7.0  7.1  7.1  7.2  7.3  7.4  7.5  7.6  7.8  7.9  7.9  7.9  8.0  8.5     6.9

Anyang    6  104  150  160  179  184  189  214  216  287  298  315  373  406  410    411  413  429  460  480  484  485  490  524  551  630  637  650  670  816     387
        4.1  5.7  6.1  6.1  6.2  6.2  6.2  6.3  6.3  6.6  6.7  6.7  6.9  7.0  7.0    7.0  7.1  7.1  7.2  7.3  7.4  7.4  7.4  7.5  7.6  7.9  7.9  8.0  8.1  9.6     7.0

Austra    4   22   61   69  100  102  106  178  248  263  281  307  355  356  375    383  391  403  517  552  554  621  624  628  636  662  664  768  783  830     395
        4.0  4.8  5.4  5.4  5.7  5.7  5.7  6.2  6.4  6.5  6.6  6.7  6.9  6.9  6.9    7.0  7.0  7.0  7.5  7.6  7.6  7.8  7.9  7.9  7.9  8.0  8.0  8.7  8.9   10     7.0

S.Cruz   18  131  171  192  208  246  252  265  278  283  302  354  400  405  408    496  498  521  575  584  615  617  620  639  646  684  697  704  728  742     443
        4.6  5.9  6.2  6.3  6.3  6.4  6.5  6.5  6.6  6.6  6.7  6.9  7.0  7.0  7.0    7.4  7.4  7.5  7.7  7.7  7.8  7.8  7.8  7.9  8.0  8.2  8.3  8.3  8.5  8.6     7.2

Dogon    98  114  176  202  206  249  255  339  378  379  386  397  407  421  430    491  513  540  543  564  577  610  645  658  665  666  685  763  773  789     464
        5.7  5.8  6.2  6.3  6.4  6.5  6.8  6.9  7.0  7.0  7.0  7.1  7.1  7.1  7.1    7.4  7.4  7.6  7.6  7.6  7.7  7.8  8.0  8.0  8.1  8.2  8.7  8.8  9.0     7.3

Atayal   49   94  130  149  212  279  284  385  431  465  471  481  489  510  511    546  559  565  568  580  581  597  613  652  707  718  745  774  790    .     481
        5.2  5.6  5.9  6.1  6.3  6.6  6.7  7.0  7.1  7.3  7.3  7.3  7.4  7.4  7.4    7.6  7.6  7.6  7.6  7.7  7.7  7.7  7.7  7.8  8.0  8.3  8.4  8.6  8.9  9.0    .     7.3

Norse    87  161  163  190  233  251  268  295  322  380  440  451  458  494  525    567  603  654  660  673  678  682  686  693  698  708  741  754  764  800     503
        5.6  6.1  6.1  6.2  6.4  6.5  6.6  6.7  6.8  6.9  7.2  7.2  7.2  7.4  7.5    7.6  7.8  8.0  8.0  8.1  8.2  8.2  8.2  8.2  8.3  8.3  8.6  8.7  8.7  9.2     7.5

Teita   186  205  210  213  230  276  327  345  351  399  418  422  439  450  569    573  588  591  600  648  656  663  689  712  749  751  761  766  796  828     514
        6.2  6.3  6.3  6.3  6.4  6.6  6.8  6.8  6.9  7.0  7.1  7.1  7.2  7.2  7.6    7.7  7.7  7.7  7.8  8.0  8.0  8.0  8.2  8.4  8.6  8.6  8.7  8.7  9.1  9.9     7.6

Peru    107  125  270  319  332  362  374  434  448  454  457  474  519  531  533    536  549  585  594  599  601  607  608  667  690  713  727  732  757  792     517
        5.7  5.9  6.6  6.7  6.8  6.9  6.9  7.2  7.2  7.2  7.2  7.3  7.5  7.5  7.5    7.6  7.6  7.7  7.7  7.8  7.8  7.8  7.8  8.1  8.2  8.4  8.5  8.5  8.7  9.0     7.5

Andama  196  244  292  303  324  335  371  393  394  428  437  477  495  527  541    550  555  578  583  595  614  641  643  676  687  703  706  729  730  787     518
        6.3  6.4  6.7  6.8  6.8  6.9  7.0  7.0  7.1  7.2  7.3  7.4  7.5  7.6    7.6  7.6  7.7  7.7  7.7  7.8  7.9  8.0  8.2  8.2  8.3  8.3  8.5  8.5  9.0     7.5

Berg     32  172  215  310  312  329  337  338  349  436  456  508  518  522  553    560  571  582  587  629  633  649  651  722  739  748  758  788  801  836     521
        5.0  6.2  6.3  6.7  6.7  6.8  6.8  6.8  6.8  7.2  7.2  7.4  7.5  7.5  7.6    7.6  7.7  7.7  7.7  7.9  7.9  8.0  8.0  8.5  8.6  8.6  8.7  9.0  9.2   10     7.6

Morior  199  209  217  219  259  318  376  377  390  401  447  501  529  530  538    556  576  579  626  655  668  669  671  677  694  753  778  782  791  808     526
        6.3  6.3  6.3  6.3  6.5  6.7  6.9  6.9  7.0  7.0  7.2  7.4  7.5  7.5  7.6    7.6  7.7  7.7  7.9  8.0  8.1  8.1  8.1  8.2  8.2  8.6  8.8  8.9  9.0  9.3     7.6

Arikar  170  289  346  389  404  433  441  482  506  512  523  526  537  547  572    606  618  680  688  692  721  734  743  746  777  785  793  794  799  807     589
        6.2  6.6  6.8  7.0  7.0  7.2  7.2  7.3  7.4  7.4  7.5  7.5  7.6  7.6  7.7    7.8  7.8  8.2  8.2  8.2  8.4  8.5  8.6  8.6  8.8  8.9  9.0  9.1  9.2  9.3     7.9

Easter  133  183  194  229  306  320  321  342  492  605  616  623  632  674  679    710  717  719  735  738  752  760  775  776  786  797  809  817  821  823     596
        5.9  6.2  6.3  6.4  6.7  6.7  6.8  6.8  7.4  7.8  7.8  7.8  7.9  8.1  8.2    8.4  8.4  8.4  8.5  8.6  8.6  8.7  8.8  8.8  8.9  9.2  9.3  9.7  9.8  9.8     8.0

Bushma  231  240  311  325  331  395  398  453  472  473  589  592  619  701  711    714  723  724  725  731  744  762  765  769  770  772  780  781  804  831     608
        6.4  6.4  6.7  6.8  6.8  7.0  7.0  7.2  7.3  7.7  7.7  7.8  8.3  8.4    8.4  8.5  8.5  8.5  8.5  8.6  8.7  8.7  8.7  8.8  8.8  8.8  8.9  9.2   10     8.0

Buriat  203  452  461  467  509  539  590  638  657  683  709  733  736  755  779    795  802  803  810  811  813  814  815  819  825  829  833  834  837  838     706
        6.3  7.2  7.3  7.3  7.4  7.6  7.7  7.9  8.0  8.2  8.4  8.5  8.5  8.7  8.8    9.1  9.2  9.2  9.3  9.4  9.4  9.5  9.5  9.8  9.9   10   10   10   11   11     8.8

Eskimo  419  470  542  570  574  593  611  631  691  696  700  715  737  740  750    756  767  784  798  805  806  812  818  820  822  824  826  827  835  839     719
        7.1  7.3  7.6  7.7  7.7  7.7  7.8  7.9  8.2  8.3  8.3  8.4  8.6  8.6  8.6    8.7  8.7  8.9  9.2  9.2  9.3  9.4  9.7  9.8  9.9  9.9  9.9  9.9   10   11     8.8
```

T A B L E 9C-10
POPKIN results: Afalou 9 #2077

```
Zalava  1   3   5   8   9   11  13  17  19  21  23  27  29  34  41    44  47  52  79  82  86  88  105 114 115 164 173 191 277 285   72
        4.0 4.2 4.6 4.7 4.8 5.1 5.1 5.2 5.3 5.3 5.4 5.5 5.6 5.6 5.7   5.8 5.8 5.9 6.3 6.3 6.3 6.3 6.4 6.5 6.5 6.8 6.9 7.0 7.4 7.4   5.8

Tasman  6   10  22  25  28  40  49  62  63  67  69  73  80  81  101   108 111 125 141 149 166 187 196 209 225 232 250 272 303 327   126
        4.7 4.9 5.4 5.5 5.5 5.7 5.9 6.1 6.1 6.2 6.2 6.2 6.3 6.3 6.4   6.5 6.5 6.6 6.7 6.7 6.8 7.0 7.0 7.1 7.2 7.2 7.3 7.4 7.6 7.7   6.4

Berg    16  24  31  35  36  39  48  78  84  92  93  104 107 109 110   113 116 120 132 178 179 181 183 207 223 260 270 364 401 405   145
        5.1 5.5 5.6 5.6 5.7 5.7 5.9 6.3 6.4 6.4 6.4 6.4 6.5 6.5 6.5   6.5 6.5 6.5 6.6 6.9 6.9 6.9 7.1 7.2 7.4 7.4 7.8 8.0 8.0   6.6

Tolai   2   4   14  20  38  53  54  58  83  98  99  102 124 138 144   154 177 180 214 253 283 306 308 316 349 355 372 468 526 714   200
        4.2 4.5 5.1 5.3 5.7 6.0 6.0 6.1 6.3 6.4 6.4 6.4 6.6 6.7 6.7   6.7 6.9 6.9 7.1 7.3 7.4 7.6 7.6 7.6 7.8 7.8 7.8 8.3 8.5 9.4   6.8

Norse   12  30  33  64  74  87  97  106 117 118 121 130 161 162 185   210 220 259 267 275 289 292 344 363 399 459 512 532 604 659   239
        5.1 5.6 5.6 6.1 6.2 6.3 6.4 6.5 6.5 6.5 6.5 6.6 6.8 6.8 7.0   7.1 7.2 7.3 7.4 7.4 7.5 7.5 7.7 7.8 8.0 8.2 8.4 8.5 8.9 9.2   7.1

S.Cruz  60  70  72  91  140 147 165 167 171 174 194 219 227 228 238   242 248 261 287 301 318 367 377 385 428 448 461 482 494 596   269
        6.1 6.2 6.2 6.4 6.7 6.7 6.8 6.8 6.9 6.9 7.0 7.2 7.2 7.2 7.3   7.3 7.3 7.4 7.5 7.6 7.7 7.8 7.9 7.9 8.1 8.2 8.2 8.3 8.4 8.9   7.3

Philip  7   55  56  66  129 135 137 146 150 153 170 192 195 199 230   244 247 256 326 348 353 357 406 416 421 433 495 558 579 671   271
        4.7 6.0 6.0 6.2 6.6 6.6 6.7 6.7 6.7 6.7 6.8 7.0 7.0 7.0 7.2   7.3 7.3 7.3 7.7 7.8 7.8 7.8 8.0 8.0 8.1 8.1 8.4 8.6 8.7 9.2   7.3

Austra  15  18  26  42  76  95  96  134 148 155 160 200 215 229 231   239 258 279 335 338 342 361 362 378 384 516 621 644 775 818   283
        5.1 5.2 5.5 5.8 6.3 6.4 6.4 6.6 6.7 6.8 7.0 7.1 7.2 7.2 7.2   7.3 7.3 7.4 7.7 7.7 7.7 7.7 7.8 7.8 7.9 7.9 8.5 9.0 9.1 10    7.3

Peru    45  65  77  94  126 145 152 175 184 186 188 212 221 222 245   251 282 296 319 358 360 366 390 400 402 454 472 540 640 703   286
        5.8 6.1 6.3 6.4 6.6 6.7 6.7 6.9 7.0 7.0 7.0 7.1 7.2 7.2 7.3   7.3 7.4 7.5 7.7 7.8 7.8 7.8 7.9 8.0 8.0 8.2 8.3 8.6 9.1 9.4   7.4

Guam    32  51  57  133 143 156 158 234 237 255 262 297 312 315 328   331 336 343 365 386 391 411 451 466 476 479 500 529 565 608   320
        5.6 5.9 6.0 6.6 6.7 6.7 6.8 7.2 7.3 7.3 7.4 7.5 7.6 7.6 7.7   7.7 7.7 7.7 7.8 7.9 7.9 8.0 8.2 8.2 8.3 8.3 8.4 8.5 8.7 8.9   7.5

Egypt   50  71  123 127 131 172 189 197 205 224 226 235 243 264 269   288 324 340 397 408 419 422 432 455 489 496 564 580 628 683   322
        5.9 6.2 6.6 6.6 6.6 6.9 7.0 7.0 7.1 7.2 7.2 7.2 7.3 7.4 7.4   7.5 7.7 7.7 8.0 8.0 8.0 8.1 8.1 8.2 8.3 8.4 8.7 8.7 9.0 9.3   7.6

Ainu    85  89  100 103 112 202 203 204 216 217 246 263 300 313 314   325 345 381 410 429 434 437 439 464 480 519 559 593 610 648   335
        6.3 6.3 6.4 6.4 6.5 7.1 7.1 7.1 7.1 7.1 7.3 7.4 7.6 7.6 7.6   7.7 7.8 7.9 8.0 8.1 8.1 8.1 8.1 8.2 8.3 8.5 8.7 8.8 8.9 9.1   7.6

Arikar  37  61  75  193 206 211 213 218 241 252 254 257 268 302 321   322 332 369 370 383 388 392 435 488 511 513 537 541 584 776   335
        5.7 6.1 6.2 7.0 7.1 7.1 7.1 7.2 7.3 7.3 7.3 7.3 7.4 7.6 7.7   7.7 7.7 7.8 7.8 7.9 7.9 7.9 8.1 8.3 8.4 8.4 8.5 8.6 8.8 8.9   7.6

NJapan  119 139 151 157 249 271 273 280 281 291 304 307 339 356 359   375 407 467 492 507 515 525 531 548 558 588 597 649 665 801   404
        6.5 6.7 6.7 6.7 7.3 7.4 7.4 7.4 7.4 7.5 7.6 7.6 7.7 7.8 7.8   7.9 8.0 8.2 8.4 8.4 8.5 8.5 8.5 8.6 8.6 8.8 8.9 9.1 9.3 10    8.0

Hainan  122 159 163 190 198 265 276 309 311 317 350 354 373 382 393   395 415 477 478 493 514 557 567 583 590 680 682 713 732 771   427
        6.5 6.8 6.8 7.0 7.0 7.4 7.4 7.6 7.6 7.6 7.8 7.8 7.8 7.9 7.9   8.0 8.0 8.3 8.3 8.4 8.4 8.6 8.7 8.8 8.9 9.3 9.3 9.4 9.6 10    8.1

Atayal  43  136 176 208 236 284 290 294 329 389 417 427 469 503 505   508 517 522 528 533 563 572 577 578 625 652 681 754 762   .     451
        5.8 6.6 6.9 7.1 7.2 7.4 7.5 7.5 7.7 7.9 8.0 8.1 8.3 8.4 8.4   8.4 8.5 8.5 8.5 8.5 8.7 8.7 8.7 8.7 9.0 9.1 9.3 9.8 9.9   .     8.2

Zulu    68  266 278 293 298 299 341 346 352 371 430 444 463 483 490   501 502 527 609 629 637 639 670 686 717 735 768 773 795 806   507
        6.2 7.4 7.4 7.5 7.5 7.5 7.7 7.8 7.8 7.8 8.1 8.1 8.2 8.3 8.3   8.4 8.4 8.5 8.9 9.0 9.1 9.1 9.2 9.3 9.5 9.6 10  10  10  10    8.5

Hawaii  168 274 286 295 380 409 418 420 426 460 474 484 487 491 521   530 535 536 547 566 619 620 634 638 676 706 707 718 734 789   518
        6.8 7.4 7.4 7.5 7.9 8.0 8.0 8.1 8.1 8.2 8.3 8.3 8.3 8.4 8.5   8.5 8.5 8.5 8.6 8.7 9.0 9.0 9.1 9.1 9.2 9.4 9.4 9.5 9.6 10    8.5

SJapan  182 310 333 334 387 403 443 447 456 471 497 524 525 553 560   568 576 587 595 602 603 627 636 675 741 743 744 753 785 836   550
        6.9 7.6 7.7 7.7 7.9 8.0 8.1 8.1 8.2 8.3 8.4 8.5 8.5 8.6 8.7   8.7 8.7 8.8 8.9 8.9 8.9 9.0 9.1 9.2 9.7 9.7 9.7 9.8 10  12    8.7

Anyang  59  90  142 374 412 413 424 457 506 544 546 592 605 606 614   632 645 658 664 690 692 700 701 709 710 724 725 739 758 821   565
        6.1 6.3 6.7 7.9 8.0 8.0 8.1 8.2 8.4 8.6 8.6 8.8 8.9 8.9 9.0   9.0 9.1 9.2 9.2 9.3 9.3 9.4 9.4 9.4 9.4 9.5 9.5 9.7 9.9 11    8.7

Eskimo  330 347 376 379 398 404 425 438 441 475 498 499 504 582 591   598 611 615 650 656 708 715 733 761 763 764 774 784 803 830   582
        7.7 7.8 7.9 7.9 7.9 8.0 8.0 8.1 8.1 8.1 8.3 8.4 8.4 8.4 8.8   8.8 8.9 8.9 9.0 9.1 9.2 9.4 9.5 9.6 9.9 9.9 9.9 10  10  11    8.9

Buriat  46  128 169 233 305 394 453 542 552 555 562 601 613 622 631   647 662 679 698 702 704 719 720 738 796 800 809 810 824 831   585
        5.8 6.6 6.8 7.2 7.6 7.9 8.2 8.6 8.6 8.6 8.7 8.9 8.9 9.0 9.0   9.1 9.2 9.2 9.4 9.4 9.4 9.5 9.5 9.6 10  10  10  11  11  11    9.0

Andama  201 323 337 414 445 452 481 510 520 538 570 575 589 630 633   646 651 660 665 666 672 678 689 696 705 731 752 759 766 823   593
        7.0 7.7 7.7 8.0 8.1 8.2 8.3 8.4 8.5 8.5 8.7 8.7 8.8 9.0 9.1   9.1 9.1 9.2 9.2 9.2 9.2 9.2 9.3 9.4 9.4 9.6 9.8 9.9 10  11    8.9

Easter  368 396 423 431 440 442 486 509 543 594 600 624 663 674 684   688 694 722 726 727 737 745 760 782 788 791 794 798 812 833   642
        7.8 8.0 8.1 8.1 8.1 8.1 8.3 8.4 8.6 8.9 8.9 9.0 9.2 9.2 9.3   9.3 9.3 9.5 9.5 9.5 9.6 9.7 9.9 10  10  10  10  10  11  11    9.3

Morior  320 449 465 473 534 539 554 571 586 607 612 635 677 687 693   695 697 716 728 740 746 756 757 778 786 792 807 814 817 832   662
        7.7 8.2 8.2 8.3 8.5 8.6 8.6 8.7 8.8 8.9 8.9 9.1 9.2 9.3 9.3   9.4 9.4 9.5 9.5 9.7 9.7 9.9 9.9 10  10  10  10  11  11  11    9.4

Teita   351 450 458 518 549 550 551 574 599 616 626 642 643 657 669   673 691 729 742 747 750 751 770 783 790 805 816 822 825 837   666
        7.8 8.2 8.2 8.5 8.6 8.6 8.6 8.7 8.9 9.0 9.0 9.1 9.1 9.2 9.2   9.2 9.3 9.5 9.7 9.8 9.8 9.8 10  10  10  10  11  11  11  12    9.4

Bushma  240 446 573 581 585 617 618 653 654 661 668 699 711 712 721   723 730 736 749 769 779 781 787 793 797 804 808 819 829 838   696
        7.3 8.1 8.7 8.7 8.9 9.0 9.0 9.1 9.2 9.2 9.2 9.4 9.4 9.4 9.5   9.5 9.6 9.6 9.8 10  10  10  10  10  10  10  10  11  11  12    9.6

Dogon   436 462 470 485 545 561 569 623 641 655 667 748 755 765 767   772 777 780 799 802 811 813 815 820 826 827 828 834 835 839   711
        8.1 8.2 8.3 8.3 8.6 8.7 8.7 9.0 9.1 9.2 9.2 9.8 9.8 9.9 10    10  10  10  10  10  11  11  11  11  11  11  11  11  12  12    9.9
```

TABLE 9C-11
POPKIN results: Teviec 11 #2078

Easter	1	2	3	4	6	7	9	11	12	13	18	27	37	41	45		46	58	61	71	93	96	98	112	121	127	156	188	199	242	255		72
	4.0	4.1	4.2	4.4	4.5	4.5	4.6	4.8	4.8	4.8	5.1	5.4	5.5	5.6	5.6		5.6	5.8	5.8	5.9	6.2	6.2	6.2	6.4	6.4	6.4	6.6	6.8	6.9	7.1	7.1		5.6
Hawaii	5	8	14	16	21	24	28	34	62	63	75	79	94	100	101		102	104	108	115	125	131	135	137	165	184	213	224	231	244	346		110
	4.5	4.6	4.8	5.0	5.2	5.4	5.4	5.5	5.5	5.9	5.9	6.0	6.0	6.2	6.3		6.3	6.3	6.3	6.4	6.4	6.5	6.5	6.5	6.7	6.8	6.9	7.0	7.0	7.1	7.6		6.1
Guam	15	20	25	30	42	48	49	72	73	80	85	88	95	118	151		178	185	233	240	265	267	302	303	316	317	324	325	407	462	484		187
	4.9	5.1	5.4	5.5	5.6	5.6	5.7	5.9	6.0	6.0	6.1	6.1	6.2	6.4	6.6		6.8	6.8	7.1	7.1	7.2	7.2	7.4	7.4	7.4	7.4	7.4	7.4	7.8	8.0	8.1		6.6
Tolai	10	31	32	39	53	57	60	64	82	90	92	120	124	130	133		183	203	208	214	245	246	264	295	362	373	374	393	446	450	644		201
	4.6	5.5	5.5	5.6	5.7	5.8	5.8	5.9	6.0	6.2	6.2	6.4	6.4	6.5	6.5		6.8	6.9	6.9	6.9	7.1	7.1	7.2	7.3	7.6	7.7	7.7	7.8	8.0	8.0	8.8		6.7
Ainu	23	29	33	51	59	70	122	141	176	186	191	210	212	249	274		307	332	338	352	353	359	398	432	444	480	508	544	594	645	646		292
	5.3	5.4	5.5	5.7	5.8	5.9	6.4	6.6	6.8	6.8	6.8	6.9	6.9	7.1	7.3		7.4	7.5	7.5	7.6	7.6	7.6	7.8	7.9	8.0	8.1	8.2	8.4	8.5	8.8	8.8		7.2
Zalava	35	36	50	65	66	67	170	174	180	182	192	193	217	227	254		276	296	304	341	368	370	418	464	472	473	500	547	563	639	708		295
	5.5	5.5	5.7	5.9	5.9	5.9	6.7	6.7	6.8	6.8	6.8	6.9	6.9	7.0	7.1		7.3	7.3	7.4	7.5	7.7	7.7	7.9	8.0	8.1	8.1	8.2	8.4	8.5	8.8	9.2		7.2
Anyang	52	68	110	147	157	159	161	190	206	219	222	234	253	278	279		283	284	293	297	299	309	318	337	360	435	490	550	599	638	704		298
	5.7	5.9	6.3	6.6	6.6	6.7	6.7	6.8	6.9	7.0	7.0	7.1	7.1	7.3	7.3		7.3	7.3	7.3	7.4	7.4	7.4	7.4	7.5	7.6	7.9	8.2	8.4	8.6	8.8	9.1		7.3
Morior	17	43	44	69	76	86	117	148	160	167	169	179	187	196	200		209	239	285	335	343	452	517	553	572	595	596	605	643	679	713		306
	5.0	5.6	5.6	5.9	6.0	6.1	6.4	6.6	6.7	6.7	6.7	6.8	6.8	6.9	6.9		6.9	7.1	7.3	7.5	7.5	8.0	8.3	8.4	8.5	8.6	8.6	8.6	8.8	9.0	9.2		7.2
Norse	40	56	74	106	109	149	162	195	207	218	226	230	259	334	356		366	380	399	402	422	441	470	471	501	527	549	575	641	732	765		345
	5.6	5.8	6.0	6.3	6.3	6.6	6.7	6.9	6.9	7.0	7.0	7.0	7.2	7.5	7.6		7.6	7.7	7.8	7.8	7.9	8.0	8.1	8.1	8.2	8.3	8.4	8.5	8.8	9.3	9.7		7.5
Eskimo	103	107	128	136	150	153	164	172	194	241	248	261	266	270	330		349	369	385	392	405	486	493	495	498	502	571	652	659	697	707		353
	6.3	6.3	6.4	6.5	6.6	6.6	6.7	6.7	6.9	7.1	7.1	7.2	7.2	7.2	7.5		7.6	7.7	7.7	7.8	7.8	8.1	8.2	8.2	8.2	8.2	8.5	8.8	8.9	9.1	9.2		7.5
NJapan	26	55	77	87	123	144	154	205	221	237	256	269	275	290	301		328	347	367	416	438	483	505	515	584	585	637	672	716	718	770		360
	5.4	5.8	6.0	6.1	6.4	6.6	6.6	6.6	6.9	7.0	7.1	7.2	7.2	7.3	7.3		7.4	7.6	7.6	7.8	8.0	8.1	8.2	8.3	8.5	8.5	8.8	9.0	9.2	9.2	9.7		7.5
Egypt	81	113	139	152	171	198	215	252	291	294	298	311	319	339	348		361	364	388	389	391	459	463	503	522	541	551	606	626	648	700		368
	6.0	6.4	6.5	6.6	6.7	6.9	6.9	7.1	7.3	7.3	7.4	7.4	7.4	7.5	7.6		7.6	7.6	7.7	7.7	7.8	8.0	8.0	8.2	8.3	8.4	8.4	8.6	8.7	8.9	9.1		7.6
Teita	38	83	89	105	142	145	223	271	272	280	282	289	313	314	321		371	404	423	425	426	449	467	534	574	577	608	635	699	774	817		378
	5.6	6.0	6.1	6.3	6.6	6.6	7.0	7.3	7.3	7.3	7.3	7.3	7.4	7.4	7.4		7.7	7.8	7.9	7.9	7.9	8.0	8.1	8.4	8.5	8.5	8.6	8.7	9.1	9.7	11		7.7
Philip	97	99	116	143	175	189	220	263	286	288	308	327	333	336	344		354	357	377	417	458	489	512	513	532	552	633	662	666	669	696		380
	6.2	6.2	6.4	6.6	6.7	6.8	7.0	7.2	7.3	7.3	7.4	7.4	7.5	7.5	7.5		7.6	7.6	7.7	7.9	8.0	8.1	8.3	8.3	8.3	8.4	8.7	8.9	8.9	8.9	9.1		7.7
SJapan	84	91	111	129	134	138	163	168	228	268	281	300	372	379	394		401	413	447	465	466	492	506	524	548	568	631	690	720	741	789		388
	6.0	6.2	6.3	6.5	6.5	6.5	6.7	6.7	7.0	7.2	7.3	7.4	7.7	7.7	7.8		7.8	7.8	8.0	8.1	8.1	8.2	8.2	8.3	8.4	8.5	8.7	9.0	9.2	9.4	10		7.7
Zulu	54	78	140	155	173	177	260	292	326	355	381	403	409	429	448		479	510	516	561	569	570	573	582	583	604	610	660	695	714	792		437
	5.7	6.0	6.5	6.6	6.7	6.8	7.2	7.3	7.4	7.6	7.7	7.8	7.8	7.9	8.0		8.1	8.2	8.3	8.4	8.5	8.5	8.5	8.5	8.5	8.6	8.6	8.9	9.1	9.2	10		7.9
Hainan	19	119	166	181	202	216	247	331	384	400	414	424	430	433	434		451	519	520	564	565	580	592	601	602	611	634	653	675	682	747		447
	5.1	6.4	6.7	6.8	6.9	6.9	7.1	7.5	7.7	7.8	7.8	7.9	7.9	7.9	7.9		8.0	8.3	8.3	8.5	8.5	8.5	8.5	8.6	8.6	8.6	8.7	8.8	9.0	9.0	9.4		7.9
Tasman	22	47	114	197	211	225	273	287	342	351	376	437	440	443	485		540	543	555	560	567	598	607	625	636	647	651	710	738	776	782		460
	5.2	5.6	6.4	6.9	6.9	7.0	7.3	7.3	7.5	7.6	7.7	7.9	8.0	8.0	8.1		8.4	8.4	8.4	8.4	8.5	8.6	8.6	8.7	8.8	8.8	8.8	9.2	9.4	9.7	9.8		8.0
Atayal	229	305	340	358	363	365	382	408	412	436	442	454	457	476	488		514	518	523	546	562	614	616	632	676	687	692	721	731	763	.		507
	7.0	7.4	7.5	7.6	7.6	7.6	7.7	7.8	7.8	7.9	8.0	8.0	8.0	8.1	8.1		8.3	8.3	8.3	8.4	8.5	8.7	8.7	8.7	9.0	9.0	9.1	9.3	9.3	9.7	.		8.3
Peru	158	204	251	258	322	345	387	396	411	420	428	445	496	497	499		525	536	558	587	603	613	656	658	661	665	683	740	775	796	800		512
	6.7	6.9	7.1	7.2	7.4	7.6	7.7	7.8	7.8	7.9	7.9	8.0	8.2	8.2	8.2		8.3	8.4	8.4	8.5	8.6	8.6	8.9	8.9	8.9	8.9	9.0	9.4	9.7	10	10		8.3
Berg	146	232	257	262	312	320	378	383	397	419	431	439	460	468	478		511	554	622	649	670	678	689	711	715	733	757	767	805	825	837		530
	6.6	7.1	7.2	7.4	7.4	7.4	7.7	7.7	7.8	7.9	7.9	8.0	8.0	8.1	8.1		8.3	8.4	8.7	8.8	8.9	9.0	9.0	9.2	9.2	9.4	9.6	9.7	10	11	12		8.5
Arikar	201	235	243	250	415	453	474	477	491	521	530	531	539	542	556		557	559	566	593	677	701	717	730	736	746	750	758	778	787	804		564
	6.9	7.1	7.1	7.1	7.8	8.0	8.1	8.1	8.2	8.3	8.3	8.3	8.3	8.4	8.4	8.4		8.4	8.4	8.5	8.5	9.0	9.1	9.2	9.3	9.4	9.4	9.5	9.6	9.7	9.9	10	8.6
Austra	132	236	386	390	421	469	494	507	528	578	579	586	600	615	624		628	640	654	685	702	712	719	762	764	768	790	793	794	827	836		607
	6.5	7.1	7.7	7.7	7.9	8.1	8.2	8.2	8.3	8.5	8.5	8.5	8.6	8.7	8.7		8.7	8.8	8.9	9.0	9.1	9.2	9.2	9.6	9.7	9.7	10	10	10	11	12		8.8
S.Cruz	306	350	395	406	482	504	538	545	589	590	619	621	629	630	655		673	698	706	734	755	761	766	773	785	786	798	806	821	823	832		646
	7.4	7.6	7.8	7.8	8.1	8.2	8.4	8.4	8.5	8.5	8.7	8.7	8.7	8.7	8.9		9.0	9.1	9.2	9.4	9.6	9.6	9.7	9.7	9.9	9.9	10	10	11	11	11		9.1
Dogon	238	277	329	375	481	487	526	537	588	627	650	657	663	686	705		709	724	742	751	756	760	781	783	801	808	811	818	820	828	833		652
	7.1	7.3	7.5	7.7	8.1	8.1	8.3	8.4	8.5	8.7	8.8	8.9	8.9	9.0	9.1		9.2	9.3	9.4	9.5	9.6	9.6	9.8	9.8	10	10	10	11	11	11	11		9.2
Andama	315	410	427	509	529	533	591	609	612	620	664	681	684	691	694		703	723	725	726	727	735	745	748	754	759	769	771	772	784	815		661
	7.4	7.8	7.9	8.2	8.3	8.3	8.5	8.6	8.6	8.7	8.9	9.0	9.0	9.0	9.1		9.1	9.3	9.3	9.3	9.3	9.4	9.4	9.4	9.6	9.6	9.7	9.7	9.7	9.8	10		9.0
Buriat	126	310	456	576	581	623	642	668	671	680	688	693	729	749	752		753	777	780	797	802	807	810	812	814	829	830	831	834	838	839		703
	6.4	7.4	8.0	8.5	8.5	8.7	8.8	8.9	8.9	9.0	9.0	9.1	9.3	9.5	9.5		9.5	9.7	9.8	10	10	10	10	10	10	11	11	11	11	12	12		9.6
Bushma	323	455	461	475	535	597	617	618	667	674	722	728	737	739	744		745	779	788	791	795	799	803	809	813	816	819	822	824	826	835		705
	7.4	8.0	8.0	8.1	8.4	8.6	8.7	8.7	8.9	9.0	9.3	9.3	9.4	9.4	9.4		9.4	9.8	9.9	10	10	10	10	10	10	10	11	11	11	11	11		9.5

TABLE 9C-12

POPKIN results: Teviec 16 #2079

Pop																																
Zalava	3	6	9	10	17	26	35	45	55	57	82	104	106	107	122	127	139	167	202	206	223	232	243	286	291	321	382	437	455	467		165
	3.5	4.1	4.3	4.3	4.5	4.6	4.8	4.9	5.0	5.0	5.2	5.4	5.4	5.4	5.5	5.5	5.6	5.7	5.9	6.0	6.0	6.1	6.1	6.3	6.3	6.4	6.7	6.9	6.9	7.0		5.5
Ainu	2	5	8	13	18	19	34	42	52	66	74	81	89	93	95	98	113	124	187	229	262	273	288	301	340	354	368	397	491	602		171
	3.3	3.9	4.1	4.4	4.5	4.5	4.8	4.9	5.0	5.1	5.2	5.2	5.3	5.3	5.3	5.3	5.4	5.5	5.8	6.1	6.2	6.2	6.3	6.3	6.5	6.6	6.6	6.7	7.1	7.6		5.5
Philip	4	14	22	24	25	31	39	64	91	110	135	164	165	204	219	253	256	258	266	268	292	294	313	370	399	410	423	425	436	464		215
	3.7	4.4	4.6	4.6	4.6	4.8	4.9	5.1	5.3	5.4	5.6	5.7	5.7	5.9	6.0	6.2	6.2	6.2	6.2	6.2	6.3	6.3	6.4	6.6	6.7	6.8	6.8	6.8	6.8	7.0		5.8
SJapan	20	29	40	47	48	51	71	75	78	79	88	109	134	142	174	183	196	228	230	269	272	311	322	357	366	430	452	542	580	707		221
	4.5	4.7	4.9	4.9	5.0	5.0	5.1	5.2	5.2	5.2	5.3	5.4	5.6	5.6	5.8	5.8	5.9	6.1	6.1	6.2	6.2	6.4	6.4	6.6	6.6	6.8	6.9	7.3	7.5	8.2		5.9
NJapan	11	16	28	30	36	58	60	62	92	94	100	108	116	148	190	192	241	254	267	290	302	305	338	344	346	403	472	495	507	750		222
	4.4	4.4	4.7	4.7	4.8	5.1	5.1	5.1	5.3	5.3	5.3	5.4	5.5	5.7	5.8	5.9	6.1	6.2	6.2	6.3	6.3	6.4	6.5	6.5	6.6	6.7	7.0	7.1	7.2	8.6		5.9
Arikar	1	7	37	38	63	70	86	115	130	140	154	208	227	237	242	246	289	295	296	298	303	309	316	342	358	359	440	478	579	643		244
	3.0	4.1	4.8	4.9	5.1	5.1	5.2	5.5	5.5	5.6	5.7	6.0	6.1	6.1	6.1	6.1	6.3	6.3	6.3	6.3	6.4	6.4	6.4	6.5	6.6	6.6	6.9	7.0	7.5	7.8		5.9
Hainan	21	23	32	49	55	83	101	103	105	114	126	199	207	214	221	226	252	263	282	293	304	312	335	373	388	475	553	566	567	736		253
	4.5	4.6	4.8	5.0	5.0	5.2	5.3	5.4	5.4	5.5	5.5	5.9	6.0	6.0	6.0	6.0	6.2	6.2	6.3	6.3	6.4	6.4	6.5	6.7	6.7	7.0	7.4	7.4	7.4	8.4		6.0
Guam	12	61	73	85	87	102	118	119	125	138	152	158	162	176	184	203	216	250	277	306	345	362	389	407	434	443	523	533	606	648		256
	4.4	5.1	5.2	5.2	5.2	5.4	5.5	5.5	5.5	5.6	5.7	5.7	5.7	5.8	5.8	5.9	6.0	6.2	6.4	6.6	6.6	6.6	6.7	6.8	6.8	6.9	7.2	7.3	7.6	7.9		6.1
Atayal	33	54	76	99	117	137	156	177	182	189	195	198	200	247	281	287	325	329	336	351	360	378	391	408	418	454	463	468	724	.		277
	4.8	5.0	5.2	5.3	5.5	5.6	5.7	5.8	5.8	5.8	5.8	5.9	5.9	5.9	6.1	6.3	6.3	6.5	6.5	6.5	6.6	6.6	6.6	6.7	6.7	6.8	6.8	6.9	6.9	7.0	8.3 .	6.2
Norse	27	56	77	97	145	151	168	173	197	210	213	217	220	235	260	275	285	314	337	363	445	450	539	556	583	634	638	685	704	737		333
	4.7	5.0	5.2	5.3	5.6	5.7	5.7	5.8	5.9	6.0	6.0	6.0	6.0	6.1	6.2	6.2	6.3	6.4	6.5	6.6	6.9	6.9	7.3	7.4	7.5	7.8	7.8	8.1	8.2	8.4		6.5
Egypt	41	65	67	131	147	149	159	172	179	191	205	251	265	276	283	339	356	374	414	428	429	433	486	498	500	589	590	628	660	694		333
	4.9	5.1	5.1	5.5	5.7	5.7	5.7	5.8	5.8	5.9	6.0	6.2	6.2	6.2	6.3	6.5	6.6	6.7	6.8	6.8	6.8	6.8	6.7	7.1	7.1	7.1	7.5	7.5	7.7	7.9	8.1	6.4
Berg	46	59	68	96	121	128	132	155	180	193	245	255	271	341	347	349	383	385	398	409	471	510	522	527	540	571	614	621	672	691		343
	4.9	5.1	5.1	5.3	5.5	5.5	5.6	5.7	5.8	5.9	6.1	6.2	6.2	6.5	6.6	6.6	6.7	6.7	6.7	6.8	7.0	7.2	7.2	7.2	7.3	7.4	7.7	7.8	8.0	8.1		6.5
Anyang	15	44	69	141	144	163	171	201	244	259	261	270	310	326	364	367	372	379	404	419	447	461	489	535	560	594	599	603	616	642		349
	4.4	4.9	5.1	5.6	5.6	5.7	5.8	5.9	6.1	6.2	6.2	6.2	6.4	6.5	6.6	6.6	6.7	6.7	6.8	6.9	6.9	6.9	7.1	7.3	7.4	7.5	7.6	7.6	7.7	7.8		6.5
S.Cruz	43	111	112	136	150	160	161	170	212	215	218	307	308	352	376	392	416	424	427	469	470	474	485	517	536	538	613	625	734	739		363
	4.9	5.4	5.4	5.6	5.7	5.7	5.7	5.8	6.0	6.0	6.0	6.4	6.4	6.6	6.7	6.7	6.8	6.8	6.8	7.0	7.0	7.0	7.1	7.2	7.3	7.3	7.7	7.8	8.4	8.5		6.6
Peru	50	80	123	143	209	211	238	320	353	361	405	446	451	459	480	496	509	515	519	529	532	550	558	605	619	626	633	669	732	746		439
	5.0	5.2	5.5	5.6	6.0	6.0	6.1	6.4	6.6	6.6	6.8	6.9	6.9	6.9	7.0	7.1	7.2	7.2	7.2	7.3	7.3	7.3	7.4	7.6	7.7	7.7	7.8	8.0	8.4	8.6		6.9
Zulu	84	157	185	222	248	300	323	365	384	386	417	420	421	422	448	466	490	492	512	520	564	595	645	675	683	723	725	730	741	786		471
	5.2	5.7	5.8	6.0	6.1	6.3	6.5	6.6	6.7	6.7	6.8	6.8	6.8	6.8	6.9	7.0	7.1	7.1	7.2	7.2	7.4	7.5	7.8	8.0	8.1	8.3	8.4	8.4	8.5	9.1		7.1
Morior	72	133	169	181	278	327	330	332	350	380	406	435	442	465	483	488	511	547	548	574	600	631	639	673	680	726	742	763	764	774		481
	5.2	5.6	5.8	5.8	6.3	6.5	6.5	6.5	6.6	6.7	6.8	6.8	6.9	7.0	7.1	7.1	7.2	7.3	7.3	7.4	7.6	7.8	7.8	8.0	8.1	8.4	8.5	8.8	8.8	8.9		7.2
Eskimo	129	186	236	299	331	343	413	415	439	444	449	457	473	476	479	502	518	563	569	585	591	620	623	629	688	697	708	710	712	766		502
	5.5	5.8	6.1	6.3	6.5	6.5	6.8	6.8	6.9	6.9	6.9	6.9	7.0	7.0	7.0	7.1	7.2	7.4	7.5	7.5	7.7	7.7	7.8	8.1	8.2	8.3	8.3	8.3	8.8			7.2
Easter	166	188	231	239	257	274	319	355	401	411	431	453	458	551	552	559	561	570	612	622	624	637	649	658	665	682	689	749	772	808		505
	5.7	5.8	6.1	6.1	6.2	6.2	6.4	6.6	6.7	6.8	6.8	6.9	6.9	7.4	7.4	7.4	7.4	7.4	7.6	7.7	7.7	7.8	7.9	7.9	8.0	8.1	8.1	8.6	8.9	9.6		7.3
Tolai	120	146	175	225	240	249	328	348	381	441	477	481	484	497	504	521	534	543	593	596	650	654	676	679	727	759	767	778	801	828		507
	5.5	5.7	5.8	6.0	6.1	6.2	6.5	6.6	6.7	6.9	7.0	7.0	7.1	7.1	7.2	7.2	7.3	7.3	7.5	7.6	7.9	7.9	8.0	8.1	8.4	8.7	8.8	8.9	9.4	10		7.3
Hawaii	153	264	284	317	334	375	394	412	432	499	525	530	541	554	584	598	604	618	644	661	662	674	686	687	690	693	698	699	728	785		544
	5.7	6.2	6.3	6.4	6.5	6.7	6.7	6.8	6.8	7.1	7.2	7.3	7.3	7.4	7.5	7.6	7.6	7.7	7.8	7.9	8.0	8.0	8.1	8.1	8.1	8.1	8.2	8.2	8.4	9.1		7.4
Andama	178	233	318	371	400	462	493	494	506	516	544	546	572	576	581	587	609	610	611	670	677	678	702	705	713	735	754	757	775	784		572
	5.8	6.1	6.4	6.6	6.7	6.9	7.1	7.1	7.2	7.2	7.3	7.3	7.4	7.5	7.5	7.5	7.6	7.6	7.6	8.0	8.0	8.0	8.2	8.2	8.3	8.4	8.7	8.7	8.9	9.1		7.6
Teita	224	297	369	393	438	514	524	537	557	565	568	582	586	592	597	615	617	632	656	659	667	718	719	729	748	762	765	769	790	791		599
	6.0	6.3	6.6	6.7	6.9	7.2	7.2	7.3	7.4	7.4	7.4	7.5	7.5	7.5	7.6	7.7	7.7	7.8	7.9	7.9	8.0	8.3	8.4	8.6	8.8	8.8	8.9	9.1	9.1	9.1		7.7
Tasman	90	194	279	377	390	487	501	503	545	549	555	573	588	636	647	652	664	666	668	706	711	740	743	752	760	768	789	799	820	834		600
	5.3	5.9	6.3	6.7	6.7	7.1	7.1	7.2	7.3	7.3	7.4	7.4	7.5	7.8	7.9	7.9	8.0	8.0	8.0	8.2	8.3	8.5	8.5	8.6	8.8	8.8	8.9	9.1	9.4	10	11	7.9
Austra	234	280	315	395	396	426	482	505	513	562	575	601	607	608	627	640	646	655	681	701	716	745	751	753	777	809	810	813	814	835		609
	6.1	6.3	6.4	6.7	6.7	6.8	7.0	7.2	7.2	7.4	7.5	7.6	7.6	7.6	7.7	7.8	7.8	7.9	8.1	8.2	8.3	8.6	8.6	8.7	8.9	9.6	9.6	9.8	9.8	11		7.9
Buriat	333	387	402	456	460	528	578	630	641	653	671	692	695	703	714	715	720	747	761	770	780	797	798	806	812	821	825	826	827	838		680
	6.5	6.7	6.8	6.9	6.9	7.3	7.5	7.8	7.8	7.9	8.0	8.1	8.1	8.2	8.3	8.3	8.3	8.6	8.8	8.9	9.0	9.3	9.3	9.6	9.7	10	10	10	10	11		8.5
Dogon	324	531	577	635	651	657	684	696	700	709	738	756	758	771	781	783	795	796	800	802	803	805	816	818	819	824	830	833	836	837		739
	6.5	7.3	7.5	7.8	7.9	7.9	8.1	8.2	8.2	8.3	8.4	8.7	8.7	8.9	9.0	9.0	9.2	9.3	9.4	9.4	9.5	9.6	9.9	10	10	10	11	11	11	11		9.0
Bushma	508	526	663	717	721	722	731	733	744	755	773	776	779	782	787	788	792	793	794	804	807	811	815	817	822	823	829	831	832	839		764
	7.2	7.2	8.0	8.3	8.3	8.3	8.4	8.4	8.5	8.7	8.9	8.9	8.9	9.0	9.1	9.1	9.1	9.2	9.2	9.6	9.6	9.7	9.9	10	10	10	10	11	11	12		9.2

Předmostí 4 (when this is read as male). Předmostí 3 has European near neighbors, but none at a distance of less than 6.2 and Chancelade has only distant neighbors.

Such distances, or greater ones, are generated in the cases of Neanderthals, Broken Hill, etc., suggesting that any affiliations which might be read are not meaningful. Certainly, there is no indication of a special connection between Neanderthals and Europeans. This will be discussed later, in viewing the results of work by others.

Upper Cave 101, #2718, again states its problematical position, already fully discussed under DISPOP. There is no control skull with a distance as low as 5.3, but some populations are not totally remote. As with DISPOP, those least remote are Arikara, Zalavár, Easter, and Norse/Berg.

What is striking is the absence from the offing of any Asiatics, the nearest such populations being Ainu and Buriat, not the usual cluster of mainland/offshore Mongoloids. The enigma of Upper Cave 101 persists.

This skull was also tested by using the samples produced by a second random pruning (see above). The order of nearness of populations was Berg, Zalavár, Easter Island, Arikara, and Norse, slightly different from the original test (table 9) and from DISPOP. Again, however, there appear the general remoteness, indecisiveness, and equal degrees of distance among the nearer populations. This time there are three cases at distances of 5.3 or less, all European. As before, Asiatics are remote.

OTHER MEASUREMENT SETS

All work to this point has involved the full set of 57 measurements, so that the TEST skulls have included only those complete enough to furnish that set. I made no attempt, in originally gathering data, to include less complete cases, however important.

It is not difficult to use smaller numbers of measurements. Given the data on the basic series, the same programming will easily produce the framework for other target specimens. This involves the same steps: recomputing the C-scores for all individuals, developing new canonical variates, redoing all distances, and so forth. The thought of such an amount of computation would have our predecessors cold with dismay, but of course the work calls for little actual computer time in the present decade.

Liujiang

This important Chinese specimen is dated at 67,000 B.P. by present questionable estimates. It is in good condition, and in character appears entirely modern, if perhaps having supraorbitals clearly more marked than is typical for East Asia at the present. Its assigned date would make it coeval with Neanderthals of Europe and the Lev-

ant, but probably later than some quasi-moderns of the west such as the Skhūl and Qafzeh crania of Israel. It may also have been coeval with an Australoid population of Southeast Asia and Indonesia, if we accept the likelihood that such a population was poised, on what was then the Pleistocene landmass of Sundaland, for the first invasion of Australia, now believed to have taken place about 55,000 B.P. or earlier.

Accordingly, these questions would be posed regarding the Liujiang skull: how does its degree of modernity stand up under analysis like that in the preceding section, and what recent population, if any, does it approximate most closely?

Chinese scientists, following Wu Rukang, have from their first reports viewed Liujiang as a proto-Mongoloid. Coon (1962) accepted this but suggested some deviation in an Australian direction. Wolpoff, Wu, and Thorne (1984), in their long review of events in East Asia and Australia, seem to fit Liujiang into late Mongoloid evolution without much discussion (apparently the latest dating had not been circulated at the time of their publication).

Forty-seven measurements of Liujiang were made on the original specimen by me through the kindness of Professor Wu Rukang and the Institute of Vertebrate Paleontology and Paleoanthro-

TABLE 9C-13
POPKIN results: Mladeč 1 #2716

Zalava
```
 1   3   5   7   0  20  24  25  26  29  32  33  35  37  59    61  70  85  90  91 105 109 114 120 123 181 205 259 361 375   90
3.7 4.0 4.4 4.5 4.6 4.8 5.0 5.0 5.0 5.1 5.2 5.2 5.2 5.2 5.6   5.7 5.7 5.9 6.0 6.0 6.2 6.2 6.3 6.3 6.3 6.8 7.0 7.3 7.8 7.8  5.7
```

Berg
```
 9  13  14  18  28  39  47  49  52  54  63  64  66  89  94    95  98 117 130 139 142 145 187 198 200 262 268 298 313 457  125
4.6 4.6 4.6 4.8 5.1 5.3 5.4 5.5 5.6 5.6 5.7 5.7 5.7 6.0 6.0   6.0 6.1 6.3 6.4 6.5 6.5 6.5 6.8 6.9 6.9 7.3 7.3 7.4 7.5 8.2  6.1
```

Tasman
```
10  21  23  31  34  36  40  41  44  55  56  57  60  74  75    82  83  88 149 161 192 196 208 210 216 256 281 294 391 395  129
4.6 4.9 5.0 5.2 5.2 5.2 5.3 5.3 5.4 5.6 5.6 5.6 5.7 5.8 5.8   5.9 5.9 6.0 6.6 6.7 6.9 6.9 7.0 7.0 7.0 7.2 7.3 7.4 7.8 7.8  6.1
```

Norse
```
 2   6  16  19  22  38  43  46  48  73  80  86  92 100 101   121 124 158 162 165 189 191 195 215 229 232 234 293 388 680  145
3.8 4.5 4.7 4.8 4.9 5.2 5.4 5.4 5.4 5.8 5.9 6.0 6.0 6.1 6.1   6.3 6.4 6.6 6.7 6.7 6.9 6.9 6.9 7.0 7.1 7.1 7.1 7.4 7.9 9.2  6.2
```

Tolai
```
 4  11  17  27  30  42  58  62  65  67  76  93 111 113 115   116 135 137 151 154 155 159 203 212 213 233 260 411 444 543  147
4.1 4.6 4.7 5.1 5.1 5.3 5.6 5.7 5.7 5.7 5.8 6.0 6.3 6.3 6.3   6.3 6.4 6.4 6.6 6.6 6.6 6.6 6.9 7.0 7.0 7.1 7.3 7.9 8.1 8.6  6.3
```

Austra
```
12  15  45  69  81  97  99 103 125 126 131 140 143 152 153   171 175 177 184 188 211 217 241 253 276 288 410 433 647 705  202
4.6 4.7 5.4 5.7 5.9 6.1 6.1 6.2 6.4 6.4 6.4 6.5 6.5 6.6 6.6   6.8 6.8 6.8 6.8 6.8 7.0 7.0 7.1 7.2 7.3 7.4 7.9 8.0 9.1 9.3  6.7
```

Egypt
```
53  78  96 118 128 129 141 169 176 179 185 207 218 228 261   284 299 307 308 309 317 332 351 377 383 426 440 476 595 653  274
5.6 5.9 6.1 6.3 6.4 6.4 6.5 6.8 6.8 6.8 6.8 7.0 7.1 7.1 7.3   7.3 7.4 7.5 7.5 7.5 7.6 7.7 7.8 8.0 8.0 8.3 8.8 9.1      7.2
```

S.Cruz
```
84 106 119 122 132 138 147 167 168 178 186 264 266 273 277   305 339 358 387 413 419 434 443 449 475 515 533 534 557 747  318
5.9 6.2 6.3 6.3 6.4 6.5 6.5 6.7 6.7 6.8 6.8 7.3 7.3 7.3 7.3   7.5 7.6 7.7 7.8 7.9 8.0 8.0 8.0 8.1 8.3 8.4 8.5 8.5 8.6 9.6  7.4
```

Peru
```
87 108 148 180 209 219 220 227 236 240 243 278 283 289 311   340 372 378 397 407 409 423 427 429 441 472 490 519 570 661  334
6.0 6.2 6.5 6.8 7.0 7.0 7.0 7.1 7.1 7.1 7.2 7.3 7.3 7.4 7.5   7.6 7.8 7.8 7.9 7.9 7.9 8.0 8.0 8.0 8.0 8.2 8.3 8.5 8.7 9.1  7.5
```

Arikar
```
68 104 127 172 173 174 182 193 214 221 238 279 302 306 331   334 346 384 390 393 398 436 460 461 469 516 536 550 574 728  335
5.7 6.2 6.4 6.8 6.8 6.8 6.8 6.9 7.0 7.0 7.1 7.3 7.4 7.5 7.6   7.6 7.7 7.8 7.8 7.8 7.9 8.0 8.2 8.2 8.2 8.4 8.5 8.6 8.7 9.5  7.5
```

Teita
```
72  79 102 112 160 183 204 206 225 239 249 254 265 290 291   316 355 367 373 389 392 438 483 487 545 575 601 637 669 798  345
5.8 5.9 6.1 6.3 6.7 6.8 7.0 7.0 7.1 7.1 7.2 7.2 7.3 7.4 7.4   7.5 7.7 7.8 7.8 7.8 7.8 8.0 8.3 8.3 8.6 8.7 8.9 9.0 9.1  10  7.6
```

Zulu
```
71  77 136 190 247 255 269 300 303 320 324 356 359 371 481   484 485 499 501 513 527 541 599 602 603 636 646 697 703 753  428
5.8 5.9 6.4 6.9 7.2 7.2 7.3 7.4 7.5 7.5 7.6 7.7 7.7 7.8 8.3   8.3 8.3 8.4 8.4 8.5 8.6 8.8 8.9 8.9 9.0 9.1 9.3 9.3 9.7 8.0
```

Ainu
```
110 197 242 246 250 252 295 338 341 342 354 369 381 402 415   420 450 480 512 520 531 546 566 572 592 623 643 684 730 764  439
6.3 6.9 7.2 7.2 7.2 7.2 7.4 7.6 7.6 7.6 7.7 7.8 7.9 7.9 7.9   8.0 8.1 8.3 8.4 8.5 8.5 8.6 8.7 8.7 8.8 9.0 9.1 9.2 9.5 9.8  8.1
```

Bushma
```
51 194 230 275 280 286 292 312 310 313 3EF 313 34E 349 369   396 404 466 517 573 588 617 619 630 654 666 714 719 729 805  445
5.6 6.9 7.1 7.3 7.3 7.4 7.4 7.5 7.5 7.5 7.6 7.7 7.7 7.7 7.7   7.8 7.9 8.2 8.4 8.7 8.8 8.9 8.9 9.0 9.1 9.1 9.4 9.4 9.0  10  8.1
```

Easter
```
156 157 202 222 297 315 362 363 366 396 399 400 425 428 445   455 456 486 551 561 563 565 584 604 633 635 640 641 708 716  458
6.6 6.6 6.9 7.0 7.4 7.5 7.8 7.8 7.8 7.8 7.9 7.9 8.0 8.0 8.1   8.1 8.2 8.3 8.6 8.6 8.7 8.7 8.8 8.9 9.0 9.0 9.0 9.0 9.4 9.4  8.2
```

Atayal
```
164 166 223 235 245 314 323 326 368 376 403 446 448 459 465   511 521 532 538 552 553 582 614 651 672 679 702 776 778   .  470
6.7 6.7 7.1 7.1 7.2 7.5 7.6 7.6 7.8 7.8 7.9 8.1 8.1 8.2 8.2   8.4 8.5 8.5 8.6 8.6 8.6 8.7 8.9 9.1 9.2 9.2 9.3 9.9 9.9   .  8.2
```

Eskimo
```
134 144 150 224 282 285 310 333 364 416 417 422 462 473 477   488 489 492 500 571 598 613 616 657 668 686 725 733 750 766  475
6.4 6.5 6.6 7.1 7.3 7.3 7.6 7.9 7.9 7.9 8.0 8.2 8.2 8.2 8.3   8.3 8.3 8.3 8.4 8.7 8.8 8.9 8.9 9.1 9.1 9.2 9.5 9.7 9.8 8.2
```

Philip
```
170 199 237 258 287 322 329 374 430 431 442 478 482 504 507   508 547 548 558 559 590 615 631 673 681 683 695 722 748 823  501
6.8 6.9 7.1 7.3 7.4 7.6 7.6 7.8 8.0 8.0 8.0 8.3 8.3 8.4 8.4   8.4 8.6 8.6 8.6 8.6 8.8 8.9 9.0 9.2 9.2 9.2 9.3 9.4 9.6  11  8.4
```

Andama
```
226 231 274 296 330 337 344 357 360 365 418 424 453 470 496   505 514 522 539 544 576 677 693 707 711 752 761 772 775 803  508
7.1 7.1 7.3 7.4 7.6 7.6 7.7 7.7 7.7 7.8 8.0 8.0 8.1 8.2 8.3   8.4 8.4 8.5 8.6 8.6 8.7 9.2 9.3 9.4 9.4 9.7 9.7 9.9 9.9  10  8.4
```

Guam
```
50 201 325 328 336 350 379 380 386 412 414 439 447 452 502   523 526 586 594 622 638 660 662 670 692 710 734 736 762 765  509
5.5 6.9 7.6 7.6 7.6 7.7 7.8 7.8 7.8 7.9 7.9 8.0 8.1 8.1 8.4   8.5 8.5 8.8 8.8 9.0 9.0 9.1 9.1 9.1 9.3 9.4 9.5 9.5 9.7 9.8  8.4
```

Hawaii
```
251 263 321 352 353 405 408 464 497 498 524 587 600 605 606   608 609 627 634 648 649 655 675 688 727 740 746 749 770 822  569
7.2 7.3 7.6 7.7 7.7 7.9 7.9 8.2 8.3 8.4 8.5 8.8 8.9 8.9 8.9   8.9 8.9 9.0 9.0 9.1 9.1 9.2 9.2 9.5 9.6 9.6 9.6 9.8  11  8.7
```

NJapan
```
244 267 301 304 451 454 463 479 503 509 535 540 542 555 581   597 611 644 645 667 690 691 701 720 751 759 780 800 810 812  580
7.2 7.3 7.4 7.5 8.1 8.1 8.2 8.3 8.4 8.4 8.5 8.6 8.6 8.6 8.7   8.8 8.9 9.1 9.1 9.1 9.2 9.2 9.3 9.4 9.7 9.7 9.9  10  10  10  8.8
```

Hainan
```
257 270 382 385 401 458 468 506 528 529 549 556 568 579 607   612 620 628 642 678 687 706 717 718 741 742 781 783 817 819  591
7.3 7.3 7.8 7.8 7.9 8.2 8.2 8.4 8.5 8.5 8.6 8.6 8.7 8.7 8.9   8.9 8.9 9.0 9.1 9.2 9.2 9.3 9.4 9.4 9.6 9.6 9.9  10  10  11  8.9
```

Morior
```
248 272 435 437 467 471 493 510 525 580 585 589 596 624 625   632 663 698 712 724 731 744 755 757 779 789 794 802 807 825  622
7.2 7.3 8.0 8.0 8.2 8.2 8.3 8.4 8.5 8.7 8.8 8.8 8.8 8.9 9.0   9.0 9.1 9.3 9.4 9.4 9.5 9.6 9.7 9.7 9.9  10  10  10  10  11  9.0
```

SJapan
```
271 494 495 518 530 554 562 564 578 591 610 626 656 658 659   664 696 721 732 737 743 754 769 771 777 782 787 793 813 838  658
7.3 8.3 8.3 8.5 8.5 8.6 8.6 8.7 8.7 8.8 8.9 9.0 9.1 9.1 9.1   9.1 9.3 9.4 9.5 9.5 9.6 9.7 9.8 9.9 9.9  10  10  10  10  12  9.2
```

Anyang
```
133 163 370 421 560 569 621 652 665 671 682 699 700 704 715   726 738 739 745 756 758 763 768 773 774 786 814 816 820 830  664
6.4 6.7 7.8 8.0 8.6 8.7 8.9 9.1 9.1 9.2 9.2 9.3 9.3 9.3 9.4   9.5 9.5 9.5 9.6 9.7 9.7 9.8 9.8 9.9 9.9  10  10  10  11  11  9.3
```

Buriat
```
107 146 335 491 567 639 650 676 689 694 709 735 760 784 785   791 795 796 797 799 806 809 811 815 818 824 826 832 834 839  699
6.2 6.5 7.6 8.3 8.7 9.0 9.1 9.2 9.2 9.3 9.4 9.5 9.7  10  10    10  10  10  10  10  10  10  10  10  10  11  11  11  11  12  9.7
```

Dogon
```
347 406 432 474 537 577 583 593 618 629 674 685 713 723 767   788 790 792 801 804 808 821 827 828 829 831 833 835 836 837  701
7.7 7.9 8.0 8.2 8.5 8.7 8.8 8.8 8.9 9.0 9.2 9.2 9.4 9.4 9.8    10  10  10  10  10  10  11  11  11  11  11  11  12  12  12  9.8
```

TABLE 9C-14
POPKIN results: Předmostí 3 #1706

```
Norse    2   3   5  15  16  19  20  29  31  36  38  43  62  65  72    76  80  89 102 104 109 138 139 146 167 172 195 355 367 434   104
        6.5 6.8 6.8 7.3 7.3 7.4 7.4 7.6 7.7 7.7 7.7 7.8 8.1 8.2 8.2   8.3 8.3 8.4 8.4 8.5 8.5 8.7 8.7 8.8 8.9 9.0 9.1 9.9  10  10   8.2

Zalava   4   6   7   9  14  21  32  37  44  45  46  52  53  77  88    95 116 119 120 123 129 164 180 227 242 288 292 321 378 479   127
        6.8 6.9 7.0 7.1 7.3 7.4 7.7 7.7 7.8 7.9 7.9 8.0 8.0 8.3 8.4   8.4 8.6 8.6 8.6 8.6 8.7 8.9 9.0 9.3 9.4 9.6 9.7 9.8  10  10   8.4

Morior  11  17  24  28  42  49  66  73  75  83 127 133 134 136 140   144 145 168 176 193 198 218 220 238 265 293 332 336 356 468   163
        7.2 7.4 7.5 7.6 7.8 7.9 8.2 8.3 8.3 8.3 8.7 8.7 8.7 8.7 8.7   8.8 8.8 8.9 9.0 9.1 9.1 9.2 9.2 9.4 9.5 9.7 9.8 9.8 9.9  10   8.8

Egypt    1  22  25  48  56  57  58  63  70  99 114 124 147 149 155   156 161 183 189 194 202 217 225 234 261 309 327 348 478 544   174
        6.2 7.4 7.5 7.9 8.0 8.0 8.0 8.1 8.2 8.4 8.6 8.6 8.8 8.8 8.9   8.9 8.9 9.0 9.1 9.1 9.1 9.2 9.3 9.3 9.5 9.7 9.8 9.9  10  11   8.8

Berg    27  33  34  35  40  41  47  54  59  64  74  87 141 158 165   179 181 197 204 231 243 267 303 368 381 389 514 551 558 696   211
        7.6 7.7 7.7 7.7 7.8 7.8 7.9 8.0 8.0 8.1 8.3 8.3 8.7 8.9 8.9   9.0 9.0 9.1 9.1 9.3 9.4 9.5 9.7  10  10  10  11  11  11  12   9.0

Tasman   8  10  12  13  23  30  39  61  78  96  97 121 135 187 191   210 214 221 224 249 275 294 308 331 352 423 439 492 525 680   211
        7.0 7.1 7.2 7.2 7.4 7.6 7.7 8.1 8.3 8.4 8.6 8.7 9.1 9.1       9.2 9.2 9.3 9.3 9.4 9.6 9.7 9.7 9.8 9.9  10  10  10  11  12   8.9

Tolai   26  51  68  84  93 115 125 130 157 162 188 206 216 239 246   256 257 268 273 279 301 310 326 342 438 454 555 569 597 600   264
        7.5 8.0 8.2 8.3 8.4 8.6 8.7 8.7 8.9 8.9 9.1 9.2 9.2 9.4 9.4   9.5 9.5 9.5 9.6 9.6 9.7 9.7 9.8 9.9  10  10  11  11  11  11   9.4

Arikar  55  91  94 107 131 177 190 207 219 229 233 244 264 278 313   316 319 341 343 354 400 402 409 436 448 450 466 473 488 536   300
        8.0 8.4 8.4 8.5 8.7 9.0 9.1 9.2 9.2 9.3 9.3 9.4 9.5 9.6 9.7   9.8 9.8 9.9 9.9 9.9 9.9  10  10  10  10  10  10  10  10  11   9.6

S.Cruz  82  85  92 112 142 148 154 182 199 200 208 211 235 251 280   317 333 361 365 377 404 422 428 430 435 467 486 559 573 677   307
        8.3 8.3 8.4 8.5 8.7 8.8 8.9 9.0 9.1 9.1 9.2 9.2 9.4 9.5 9.6   9.8 9.8 9.9  10  10  10  10  10  10  10  10  11  11  12   9.6

Austra  50  67  98 100 103 106 111 113 122 170 205 245 324 335 340   351 353 376 411 414 424 446 455 460 503 505 556 567 683 738   324
        8.0 8.2 8.4 8.4 8.4 8.5 8.5 8.5 8.6 9.0 9.1 9.4 9.8 9.8 9.8   9.9 9.9  10  10  10  10  10  10  10  11  11  11  11  12  12   9.7

Hawaii  81 108 126 173 178 192 203 209 212 215 228 236 237 284 318   349 372 385 391 413 437 444 464 469 504 521 542 563 572 659   336
        8.3 8.5 8.7 9.0 9.0 9.1 9.1 9.2 9.2 9.3 9.4 9.4 9.6 9.8       9.9  10  10  10  10  10  11  11  11  11  11  11  11  11   9.8

Ainu   105 137 150 151 159 169 185 186 222 255 271 286 296 305 329   337 373 379 394 416 418 477 506 539 553 581 699 701 718 743   368
        8.5 8.7 8.8 8.8 8.9 8.9 9.0 9.1 9.3 9.5 9.6 9.6 9.7 9.7 9.8   9.8  10  10  10  10  10  11  11  11  11  12  12  12  12   9.9

Eskimo  71 110 128 153 163 250 285 299 300 312 325 374 375 383 388   399 407 433 483 496 498 502 538 543 547 554 578 583 748 757   399
        8.2 8.5 8.7 8.9 8.9 9.4 9.6 9.7 9.7 9.7 9.8  10  10  10  10   10  10  10  10  11  11  11  11  11  11  11  11  12  12  10.1

Teita   69 160 196 226 240 258 259 282 307 311 338 344 397 410 420   447 461 482 493 518 520 523 540 592 660 685 687 725 763 820   438
        8.2 8.9 9.1 9.3 9.4 9.5 9.5 9.6 9.7 9.7 9.8 9.9  10  10  10   10  10  10  10  11  11  11  11  11  11  12  12  12  12  13  10.3

Easter 143 166 184 248 260 263 281 306 314 346 357 384 386 401 451   462 472 474 494 534 537 545 599 614 635 676 692 733 774 792   451
        8.7 8.9 9.0 9.4 9.5 9.5 9.6 9.7 9.8 9.9 9.9  10  10  10  10   10  10  10  11  11  11  11  11  11  11  12  12  12  12  10.4

Peru   241 252 269 272 304 320 339 345 350 396 412 415 427 456 459   475 500 509 512 524 530 570 584 589 604 617 630 640 684 688   464
        9.4 9.5 9.5 9.6 9.7 9.8 9.8 9.9 9.9  10  10  10  10  10  10   11  11  11  11  11  11  11  11  11  11  11  11  12  12  10.4

NJapan 223 247 283 289 295 347 360 362 366 369 380 382 387 395 403   417 419 445 480 485 575 609 618 622 679 700 716 728 751 793   471
        9.3 9.4 9.6 9.6 9.7 9.9 9.9 9.9 9.9  10  10  10  10  10  10   10  10  10  11  11  11  11  12  12  12  12  12  12  10.5

Guam    90 171 232 253 262 290 315 323 334 359 406 440 441 449 458   507 541 557 580 615 626 647 648 663 670 690 706 714 746 784   484
        8.4 9.0 9.3 9.5 9.5 9.6 9.8 9.8 9.8 9.9  10  10  10  10  11   11  11  11  11  11  11  11  11  12  12  12  12  12  10.5

Zulu    60 118 132 152 254 291 297 358 393 431 489 499 501 511 517   519 560 562 566 582 598 611 613 652 672 703 709 776 785 801   490
        8.1 8.6 8.7 8.8 9.5 9.7 9.7 9.9  10  10  10  11  11  11  11   11  11  11  11  11  11  12  12  12  12  12  10.5

Philip  86 201 277 302 322 328 363 371 408 425 426 453 471 490 510   513 527 587 593 641 675 694 702 719 735 750 756 778 782 803   523
        8.3 9.1 9.6 9.7 9.8 9.8 9.9  10  10  10  10  10  10  11  11   11  11  12  12  12  12  12  12  12  12  12  12  12  10.7

Buriat  18 101 266 370 390 392 429 465 476 481 487 497 508 526 532   546 586 602 620 631 633 645 646 649 664 704 768 770 815 819   535
        7.4 8.4 9.1  10  10  10  10  10  11  11  11  11  11  11  11   11  11  11  11  12  12  12  12  13  13  10.8

SJapan 174 287 330 405 432 443 457 463 528 531 550 574 576 607 608   610 624 628 634 644 662 708 710 726 731 732 789 798 812 838   587
        9.0 9.6 9.8  10  10  10  10  11  11  11  11  11  11  11  11   11  11  12  12  12  12  12  12  13  14  11.1

Anyang  79 175 421 442 452 552 561 564 619 625 643 661 666 667 671   691 715 720 721 722 727 729 737 744 760 762 779 788 794 795   633
        8.3 9.0  10  10  10  11  11  11  11  11  11  11  11  11  12   12  12  12  12  12  12  12  12  12  12  12  12  12  11.3

Andama 230 298 495 522 529 535 568 571 577 579 594 595 627 632 637   656 657 681 698 711 724 754 764 790 796 797 804 811 822 824   643
        9.3 9.7  11  11  11  11  11  11  11  11  11  11  11  11  11   12  12  12  12  12  12  12  12  13  13  13  11.4

Hainan 213 274 276 364 470 491 591 601 612 621 650 668 669 674 705   707 713 717 723 741 752 753 759 767 773 799 806 810 817 818   644
        9.2 9.6 9.6 9.9  10  10  11  11  11  11  11  11  11  12  12   12  12  12  12  12  12  12  12  13  13  11.4

Bushma 117 270 398 484 515 516 549 596 603 616 623 638 642 651 655   665 686 736 745 765 781 783 800 802 805 809 816 830 831 834   652
        8.6 9.6  10  11  11  11  11  11  11  11  11  11  11  11  12   12  12  12  12  12  12  13  13  13  14  11.5

Atayal 533 565 588 606 629 636 639 653 654 658 678 689 693 695 697   712 739 761 769 771 772 775 786 787 807 808 814 823 835   .   709
        11  11  11  11  11  11  11  11  11  12  12  12  12  12       12  12  12  12  12  12  12  12  12  13  13  14   .  11.8

Dogon  548 585 590 605 673 682 730 734 740 742 747 749 755 758 766   777 780 791 813 821 825 826 827 828 829 832 833 836 837 839   757
        11  11  11  11  12  12  12  12  12  12  12  12  12  12       12  12  13  13  13  13  13  13  13  14  14  14  14  15  12.4
```

TABLE 9C-15
POPKIN results: Předmostí 4 #1707

```
Zalava   4    8    9   11   12   13   15   16   17   27   35   39   45   50   54    57   60   61   66   73   78   83   84   85  115  123  141  183  196  323    69
        4.7  5.0  5.0  5.0  5.1  5.2  5.3  5.3  5.3  5.7  6.0  6.1  6.1  6.2  6.3   6.3  6.4  6.4  6.5  6.6  6.6  6.7  6.7  6.7  6.9  7.0  7.2  7.5  7.6  8.3   6.2

Norse    1    2   10   14   19   22   23   25   28   29   36   48   56   58   63    69   90  100  107  109  120  121  130  131  166  167  168  230  325  474    98
        4.3  4.5  5.0  5.3  5.4  5.6  5.7  5.7  5.7  5.8  6.0  6.2  6.3  6.3  6.4   6.5  6.8  6.8  6.9  6.9  7.0  7.0  7.1  7.1  7.4  7.4  7.4  7.8  8.4  9.0   6.5

Egypt    3   21   30   34   40   42   53   55   59   71   74   79   82   86   89    91   94   95   96   99  104  111  114  135  153  193  199  215  272  281   102
        4.7  5.5  5.8  5.9  6.1  6.1  6.2  6.3  6.4  6.6  6.6  6.6  6.6  6.7  6.7   6.8  6.8  6.8  6.8  6.8  6.9  6.9  6.9  7.2  7.3  7.6  7.6  7.7  8.1  8.1   6.7

Tasman  24   26   43   46   67   68   98  116  122  140  162  163  165  169  172   173  181  204  232  233  236  256  258  259  263  277  289  346  394  413   187
        5.7  5.7  6.1  6.1  6.5  6.5  6.8  7.0  7.0  7.2  7.3  7.4  7.4  7.4  7.5   7.5  7.5  7.7  7.8  7.8  7.8  7.9  7.9  8.0  8.0  8.1  8.1  8.4  8.7  8.8   7.4

Tolai    6    7   33   62   77  101  105  119  126  129  143  148  154  160  171   175  203  219  220  226  227  245  248  252  336  361  378  388  448  580   202
        4.8  5.0  5.9  6.4  6.6  6.9  6.9  7.0  7.1  7.1  7.2  7.3  7.3  7.3  7.4   7.5  7.7  7.7  7.7  7.7  7.8  7.8  7.9  7.9  7.9  8.4  8.5  8.6  8.6  8.9  9.4 7.4

Berg    18   32   44   52   70   76   80  102  110  125  128  170  174  177  185   202  205  213  223  224  254  306  312  327  338  343  345  471  497  688   216
        5.4  5.9  6.1  6.2  6.6  6.6  6.6  6.9  6.9  7.1  7.1  7.4  7.5  7.5  7.5   7.6  7.7  7.7  7.8  7.8  7.9  8.3  8.3  8.4  8.4  8.4  8.4  8.9  9.1  9.9   7.5

Ainu    31   38   65   81   87  106  112  132  137  138  139  142  144  161  179   180  200  239  264  274  284  295  299  302  357  371  406  447  470  664   225
        5.9  6.1  6.4  6.7  6.7  6.9  6.9  7.1  7.2  7.2  7.2  7.2  7.2  7.3  7.5   7.5  7.6  7.8  8.0  8.1  8.1  8.2  8.2  8.2  8.5  8.6  8.7  8.9  9.0  9.8   7.6

Austra   5   37   41   47   49   51   88   97  124  133  134  146  151  178  186   187  189  197  201  218  250  334  348  377  390  408  415  432  669  682   225
        4.8  6.1  6.1  6.2  6.2  6.2  6.7  6.8  7.0  7.2  7.2  7.3  7.3  7.5  7.5   7.5  7.6  7.6  7.6  7.7  7.9  8.4  8.4  8.6  8.6  8.7  8.8  8.8  9.8  9.9   7.5

Zulu    20  108  117  127  147  152  176  184  188  206  221  255  257  269  285   286  322  333  369  397  399  402  418  445  461  502  562  587  605  632   314
        5.5  6.9  7.0  7.1  7.3  7.3  7.5  7.5  7.6  7.7  7.7  7.9  7.9  8.0  8.1   8.1  8.3  8.4  8.5  8.7  8.7  8.7  8.8  8.9  8.9  9.1  9.4  9.5  9.5  9.6   8.1

Teita   75  118  145  157  158  164  212  225  240  244  265  268  319  324  326   354  398  424  441  457  500  505  513  526  591  601  610  765  788  803   384
        6.6  7.0  7.2  7.3  7.3  7.4  7.7  7.8  7.8  7.9  8.0  8.0  8.3  8.4  8.4   8.5  8.7  8.8  8.9  8.9  9.1  9.1  9.1  9.2  9.5  9.5  9.6   10   11   11   8.5

Philip  64  192  195  198  208  214  247  249  251  262  271  311  316  331  367   387  401  411  433  456  478  506  540  544  552  613  644  646  677  719   389
        6.4  7.6  7.6  7.7  7.7  7.7  7.9  7.9  8.0  8.1  8.3  8.3  8.4  8.5  8.6   8.7  8.7  8.8  8.9  9.0  9.1  9.3  9.3  9.3  9.6  9.7  9.7  9.9  9.9   10   8.6

Peru   155  182  209  253  298  304  308  320  329  364  370  373  386  404  405   412  419  443  459  476  477  518  530  547  555  594  598  720  729  749   430
        7.3  7.5  7.7  7.9  8.2  8.2  8.3  8.3  8.4  8.5  8.6  8.6  8.6  8.7  8.7   8.8  8.8  8.9  8.9  9.0  9.0  9.2  9.2  9.3  9.4  9.5  9.5   10   10   10   8.8

Guam   113  159  194  234  261  267  290  296  335  341  349  350  395  409  463   469  485  486  499  511  522  525  561  660  661  680  706  726  733  779   449
        6.9  7.3  7.6  7.8  8.0  8.0  8.2  8.2  8.4  8.4  8.4  8.5  8.7  8.7  9.0   9.0  9.0  9.0  9.1  9.1  9.2  9.2  9.4  9.8  9.8  9.9   10   10   10   10   8.8

NJapan 207  237  238  276  280  301  310  321  342  352  360  374  420  422  430   437  450  454  464  501  503  510  520  621  652  665  668  675  736  777   450
        7.7  7.8  7.8  8.1  8.1  8.2  8.3  8.3  8.4  8.5  8.5  8.6  8.8  8.8  8.8   8.8  8.9  8.9  9.0  9.1  9.1  9.1  9.2  9.6  9.7  9.8  9.8  9.8   10   10   8.9

S.Cruz 243  282  288  300  309  313  315  347  365  375  380  391  392  416  436   460  467  468  484  493  529  549  583  620  641  697  708  730  737  797   471
        7.9  8.1  8.1  8.2  8.3  8.3  8.3  8.4  8.5  8.6  8.6  8.7  8.7  8.8  8.8   8.9  9.0  9.0  9.0  9.1  9.2  9.3  9.4  9.6  9.7   10   10   10   10   11   9.0

Dogon   72  156  231  241  260  266  307  353  356  363  372  455  489  490  523   524  531  584  627  650  686  705  710  725  751  759  762  770  811  833   512
        6.6  7.3  7.8  7.8  8.0  8.0  8.3  8.5  8.5  8.6  8.8  9.0  9.0  9.2  9.2   9.3  9.5  9.6  9.7  9.9   10   10   10   10   10   10   11   12   9.2

Hainan 103  190  191  246  317  340  379  407  431  449  458  472  473  482  495   516  519  554  590  629  636  648  707  714  718  724  732  787  792  814   517
        6.9  7.6  7.6  7.9  8.3  8.4  8.6  8.7  8.8  8.9  8.9  9.0  9.0  9.0  9.1   9.2  9.2  9.4  9.5  9.6  9.7  9.7   10   10   10   10   11   11   11   11   9.2

Atayal 150  278  328  360  389  400  423  429  440  442  481  494  517  541  542   566  567  576  581  592  623  643  671  698  712  715  767  778  781    .   534
        7.3  8.1  8.4  8.5  8.6  8.7  8.8  8.8  8.8  8.9  9.0  9.1  9.2  9.3  9.3   9.4  9.4  9.4  9.4  9.5  9.6  9.7  9.8   10   10   10   10   10   11    .   9.3

Morior 210  229  283  303  314  362  381  414  444  446  452  492  535  556  560   572  585  624  626  651  672  674  676  713  727  743  761  764  776  806   545
        7.7  7.8  8.1  8.2  8.3  8.5  8.6  8.8  8.9  8.9  8.9  9.1  9.3  9.4  9.4   9.4  9.5  9.6  9.6  9.7  9.8  9.8  9.9   10   10   10   10   10   10   11   9.3

Arikar 216  235  287  337  421  427  438  466  508  521  532  545  558  574  595   596  606  609  614  618  640  647  658  667  700  711  716  721  789  807   559
        7.7  7.8  8.1  8.4  8.8  8.8  8.8  8.9  9.0  9.1  9.2  9.3  9.3  9.4  9.4   9.5  9.5  9.6  9.6  9.6  9.7  9.7  9.8  9.8   10   10   10   11   11   9.4

Anyang  93  222  297  359  417  434  435  480  488  536  564  568  577  600  604   612  634  637  653  657  659  679  684  690  703  753  755  795  816  819   574
        6.8  7.7  8.2  8.5  8.8  8.8  8.8  8.9  9.0  9.3  9.4  9.4  9.5  9.5  9.5   9.6  9.7  9.7  9.7  9.8  9.9  9.9  9.9  9.9   10   10   10   11   11   11   9.4

Bushma  92  136  275  292  385  393  425  439  475  538  559  569  578  608  617   622  630  631  654  685  695  728  740  745  746  754  791  799  812  825   575
        6.8  7.2  8.1  8.2  8.6  8.7  8.8  8.8  9.0  9.3  9.4  9.4  9.4  9.6  9.6   9.6  9.6  9.6  9.7  9.9  9.9   10   10   10   10   10   11   11   11   11   9.5

SJapan 217  330  344  351  366  382  383  462  465  491  515  527  551  570  586   611  615  616  638  656  663  704  723  731  768  773  801  822  827  837   578
        7.7  8.4  8.4  8.5  8.5  8.6  8.6  9.0  9.0  9.1  9.2  9.2  9.3  9.4  9.5   9.6  9.6  9.6  9.7  9.8  9.8   10   10   10   10   10   11   11   11   12   9.6

Easter 228  270  273  279  305  318  403  410  451  507  537  553  557  563  602   625  691  694  701  717  734  742  747  760  772  790  808  809  818  830   583
        7.8  8.1  8.1  8.1  8.2  8.3  8.7  8.7  8.9  9.1  9.3  9.3  9.4  9.4  9.5   9.6  9.9  9.9   10   10   10   10   10   10   10   11   11   11   11   12   9.6

Andama 242  291  384  426  453  498  514  533  565  571  573  575  593  607  619   639  655  673  689  692  693  702  735  738  769  783  813  823  829  832   617
        7.8  8.2  8.6  8.8  8.9  9.1  9.1  9.3  9.4  9.4  9.4  9.4  9.5  9.6  9.6   9.7  9.8  9.8  9.9  9.9  9.9   10   10   10   10   11   11   11   11   12   9.7

Eskimo 293  332  339  358  396  479  504  509  534  546  548  579  588  589  635   649  687  699  722  750  756  758  782  785  786  805  810  815  817  824   622
        8.2  8.4  8.4  8.5  8.7  9.0  9.1  9.3  9.3  9.3  9.4  9.5  9.5  9.7  9.7   9.9   10   10   10   10   11   11   11   11   11   11   11   11   11   9.7

Hawaii 355  376  428  483  496  512  539  543  550  582  597  599  603  628  642   666  681  696  709  741  744  752  763  766  784  793  794  800  826  828   643
        8.5  8.6  8.8  9.0  9.1  9.1  9.3  9.3  9.3  9.4  9.5  9.5  9.5  9.6  9.7   9.8  9.9  9.9   10   10   10   10   10   11   11   11   11   11   11   11   9.8

Buriat 149  211  294  487  528  633  645  662  670  678  683  739  748  757  771   774  775  780  796  798  802  804  820  821  831  834  835  836  838  839   695
        7.3  7.7  8.2  9.0  9.2  9.7  9.7  9.8  9.8  9.9  9.9   10   10   10   10    10   10   10   11   11   11   11   11   11   12   12   12   12   12   13  10.4
```

TABLE 9C-16
POPKIN results: Chancelade #1705

```
Egypt    1   3   4   5   7  11  17  22  30  35  41  42  44  46  54    57  70  71  74  77  84  89  99 104 113 114 131 139 140 224    65
        7.7 8.1 8.1 8.1 8.3 8.4 8.6 8.8 9.0 9.1 9.2 9.3 9.3 9.3 9.4   9.5 9.7 9.7 9.7 9.7 9.7 9.8 9.9 9.9  10  10  10  10  10  11   9.3

Norse    2  12  15  18  26  31  32  33  37  40  47  50  55  56  61    92  93 110 112 125 129 133 158 164 166 175 190 200 232 597   106
        8.1 8.4 8.5 8.6 8.9 9.0 9.1 9.1 9.1 9.2 9.3 9.4 9.5 9.5 9.6   9.8 9.8  10  10  10  10  10  10  10  10  10  11  11  11  12   9.7

Teita    8   9  23  25  34  39  43  45  60  73  83  94 103 105 111   123 152 163 167 168 179 185 233 238 254 314 347 365 384 697   161
        8.3 8.3 8.8 8.9 9.1 9.2 9.3 9.3 9.5 9.7 9.7 9.8 9.9 9.9  10    10  10  10  10  10  11  11  11  11  11  11  11  11  12  13  10.1

Zalava  10  13  16  19  27  29  53  59  62  80  96 141 153 159 161   165 171 172 173 189 204 206 328 346 378 390 422 449 464 533   189
        8.4 8.5 8.5 8.7 9.0 9.0 9.4 9.5 9.6 9.7 9.9  10  10  10  10    10  10  10  10  11  11  11  11  11  12  12  12  12  12  12  10.3

Easter  20  24  28  36  51  78  86 108 122 127 137 145 151 183 216   222 229 295 300 304 321 329 331 339 349 353 385 459 572 672   232
        8.8 8.9 9.0 9.1 9.4 9.7 9.8 9.9  10  10  10  10  10  11  11    11  11  11  11  11  11  11  11  11  11  11  12  12  12  13  10.6

Tolai   38  52  63  82  90  95 117 147 149 154 169 176 191 196 201   220 243 252 276 313 322 325 332 374 392 403 415 443 466 496   240
        9.2 9.4 9.6 9.7 9.8 9.9  10  10  10  10  10  11  11  11  11    11  11  11  11  11  11  11  11  12  12  12  12  12  12  12  10.7

Eskimo   6  58  85  98 106 116 126 128 134 144 186 192 208 230 242   291 293 296 301 316 318 330 352 369 391 423 450 500 662 771   271
        8.2 9.5 9.7 9.9 9.9  10  10  10  10  11  11  11  11  11  11    11  11  11  11  11  11  11  11  12  12  12  12  13  14  10.9

Guam    48  79 101 102 107 143 148 156 170 177 178 193 219 223 225   240 282 290 363 397 404 412 447 473 474 497 507 567 669 735   301
        9.3 9.7 9.9 9.9 9.9 9.9  10  10  10  10  10  11  11  11  11    11  11  11  11  12  12  12  12  12  12  12  12  12  13  13  11.0

Ainu    14  21  67  68  76  87 132 160 198 202 244 265 267 308 338   343 350 356 357 359 373 417 425 435 453 505 517 549 596 608   303
        8.5 8.8 9.6 9.6 9.7 9.8  10  10  11  11  11  11  11  11  11    11  11  11  11  11  12  12  12  12  12  12  12  12  12  13  11.0

Dogon   75  81  97 119 121 155 162 180 207 227 247 250 263 280 289   292 309 315 354 360 362 366 379 387 486 621 681 700 752 792   330
        9.7 9.7 9.9  10  10  10  10  11  11  11  11  11  11  11  11    11  11  11  11  11  11  12  12  13  13  13  13  14  14  11.2

Zulu    64  66  72  91 109 135 150 181 211 235 255 258 277 283 284   358 377 381 413 451 456 475 489 494 501 555 558 588 618 626   334
        9.6 9.6 9.7 9.8 9.9  10  10  11  11  11  11  11  11  11  11    11  12  12  12  12  12  12  12  12  12  12  12  13  13  13  11.2

Hawaii  65  69 142 174 182 188 203 241 245 278 307 319 398 405 407   432 479 490 498 513 530 532 553 573 578 609 611 625 664 740   402
        9.6 9.6  10  10  11  11  11  11  11  11  11  11  12  12  12    12  12  12  12  12  12  12  12  12  13  13  13  13  13  11.6

Berg   100 115 136 184 187 199 221 259 262 269 297 320 337 342 394   400 411 442 446 467 502 512 528 531 633 648 696 772 793 824   408
        9.9  10  10  11  11  11  11  11  11  11  11  11  11  11  12    12  12  12  12  12  12  12  12  12  13  13  13  14  14  14  11.7

Peru   130 194 205 217 231 270 273 275 305 312 355 375 408 420 518   525 535 537 543 559 565 579 583 614 619 624 629 636 732 741   450
         10  11  11  11  11  11  11  11  11  11  11  12  12  12  12    12  12  12  12  12  12  12  12  13  13  13  13  13  13  11.8

Austra 120 197 266 268 274 294 311 327 344 348 351 368 383 386 396   418 438 478 495 523 542 564 585 591 604 673 688 702 780 804   451
         10  11  11  11  11  11  11  11  11  11  11  12  12  12  12    12  12  12  12  12  12  12  12  13  13  13  13  14  14  14  11.9

Morior  49  88 212 213 234 236 237 249 257 286 299 367 389 431 436   458 481 483 582 607 638 650 674 704 707 723 728 766 777 798   459
        9.4 9.8  11  11  11  11  11  11  11  11  11  11  12  12  12    12  12  12  12  13  13  13  13  13  13  13  13  14  14  14  11.9

Hainan 146 226 239 260 326 333 414 424 427 444 462 492 508 540 554   557 560 563 566 569 587 592 600 617 649 683 692 699 758 828   511
         10  11  11  11  11  11  12  12  12  12  12  12  12  12  12    12  12  12  12  12  12  13  13  13  13  13  13  13  15  12.1

SJapan 210 215 228 285 306 402 433 437 448 463 471 485 487 519 522   538 601 622 628 637 642 682 685 713 716 719 731 736 807 826   540
         11  11  11  11  11  12  12  12  12  12  12  12  12  12    12  13  13  13  13  13  13  13  13  13  13  13  13  14  15  12.3

Anyang 195 209 253 264 334 336 429 472 484 493 506 514 521 524 539   544 546 562 577 594 654 677 701 712 722 726 739 763 805 831   541
         11  11  11  11  11  11  12  12  12  12  12  12  12  12    12  12  12  12  13  13  13  13  13  13  13  14  14  15  12.3

Atayal 138 218 248 271 340 376 382 388 452 465 503 536 547 552 571   615 620 655 657 660 661 667 684 703 734 760 774 783 790   .   543
         10  11  11  11  11  12  12  12  12  12  12  12  12  12    13  13  13  13  13  13  13  13  13  14  14  14  14   .  12.3

Tasman 157 214 261 317 323 341 364 406 409 440 509 511 576 580 581   584 598 605 630 631 640 656 671 676 678 725 773 811 815 832   544
         10  11  11  11  11  11  11  12  12  12  12  12  12  12    13  13  13  13  13  13  13  13  13  13  14  14  14  15  12.3

Andama 256 345 361 370 416 428 430 434 439 454 468 476 482 488 491   570 574 589 593 602 612 645 652 665 691 694 710 791 794 808   544
         11  11  11  11  12  12  12  12  12  12  12  12  12  12    12  12  12  13  13  13  13  13  13  13  13  14  14  14  12.3

Philip 281 288 303 324 372 393 455 460 504 515 527 541 550 561 586   590 623 653 658 663 675 679 714 730 737 738 747 748 755 762   571
         11  11  11  11  11  12  12  12  12  12  12  12  12  12    12  13  13  13  13  13  13  13  13  13  13  13  13  13  14  12.4

NJapan 118 246 251 287 298 310 426 441 445 480 556 595 627 632 651   668 680 687 690 705 718 727 750 754 759 784 786 789 801 802   582
         10  11  11  11  11  12  12  12  12  12  12  13  13  13    13  13  13  13  13  13  13  13  13  14  14  14  14  14  12.5

Arikar 272 302 371 380 395 419 457 461 499 545 551 568 575 599 606   610 635 646 647 659 689 698 709 724 743 761 770 785 795 816   590
         11  11  11  12  12  12  12  12  12  12  12  13  13  13    13  13  13  13  13  13  13  14  14  14  14  14  14  14  12.5

S.Cruz 410 469 516 520 526 529 534 548 603 613 643 666 695 711 717   742 746 753 757 765 776 779 787 799 803 814 817 822 829 837   684
         12  12  12  12  12  12  12  13  13  13  13  13  13  13    13  13  13  13  14  14  14  14  14  14  14  14  15  15  13.2

Bushma 279 335 401 421 510 616 639 641 644 706 720 721 744 749 751   764 768 769 775 778 796 800 806 810 818 820 821 827 830 833   696
         11  11  12  12  12  13  13  13  13  13  13  13  13  13    14  14  14  14  14  14  14  14  14  14  14  15  15  15  13.3

Buriat 124 399 470 477 634 670 686 693 708 715 729 733 745 756 767   781 782 788 797 809 812 813 819 823 825 834 835 836 838 839   718
         10  12  12  12  13  13  13  13  13  13  13  13  13  14    14  14  14  14  14  14  14  14  14  15  15  15  15  16  16  13.6
```

TABLE 9C-17

POPKIN results: La Chapelle #2082

```
Austra    4    6   10   15   16   19   24   25   38   39   54   63   67   70   78    80   81   86   99  100  104  132  151  185  192  211  275  279  296  321   104
         9.2  9.3  9.8  10   10   10   11   11   11   11   11   11   11   11   11    12   12   12   12   12   12   12   12   12   12   12   13   13   13   13   11.4

Tasman    1    2    3    5    7    9   17   18   28   32   40   45   46   47   48    60   71   74   93  121  129  142  143  146  162  187  199  268  474  633   108
         8.7  9.0  9.1  9.3  9.4  9.6  10   10   11   11   11   11   11   11   11    11   11   11   12   12   12   12   12   12   12   12   12   13   13   14   11.2

S.Cruz   20   26   27   37   50   76   88   92  111  112  120  125  127  139  144   150  157  160  177  190  191  216  249  251  274  281  339  351  384  507   172
          11   11   11   11   11   11   12   12   12   12   12   12   12   12   12    12   12   12   12   12   12   12   13   13   13   13   13   13   13   14   12.0

Zulu     21   22   23   30   33   41   44   49   64   77   79   87  110  130  137   141  167  168  184  197  204  214  237  363  377  429  430  442  443  588   184
          11   11   11   11   11   11   11   11   11   11   12   12   12   12   12    12   12   12   12   12   12   12   13   13   13   13   13   13   13   14   12.0

NJapan    8   12   29   35   66   69   95  109  119  138  156  161  180  189  223   231  256  271  276  284  310  331  367  388  409  418  459  497  541  603   241
         9.5  10   11   11   11   11   12   12   12   12   12   12   12   12   12    12   13   13   13   13   13   13   13   13   13   13   13   14   14   14   12.3

Morior   34   58   62   91   94  108  115  134  152  166  172  173  208  220  248   262  278  286  301  303  318  324  360  402  414  453  485  501  580  599   263
          11   11   11   12   12   12   12   12   12   12   12   12   12   12   13    13   13   13   13   13   13   13   13   13   13   13   14   14   14   14   12.5

Tolai    36   43   55   68   82  103  140  148  149  158  183  200  227  228  230   280  315  328  375  378  389  390  464  467  468  477  487  498  517  521   280
          11   11   11   11   12   12   12   12   12   12   12   12   12   12   12    13   13   13   13   13   13   13   13   13   13   13   14   14   14   14   12.6

Ainu     13   53   65   73   85   97  113  114  131  196  207  210  252  259  260   288  308  313  342  348  357  385  403  416  424  436  493  546  583  702   282
          10   11   11   11   12   12   12   12   12   12   12   12   13   13   13    13   13   13   13   13   13   13   13   13   13   13   14   14   14   15   12.6

Berg     11   51   56   59   90   96  102  117  154  170  224  238  290  302  304   316  317  334  336  343  395  422  437  445  519  533  568  570  587  624   300
         9.8  11   11   11   12   12   12   12   12   12   12   12   13   13   13    13   13   13   13   13   13   13   13   13   14   14   14   14   14   14   12.6

Arikar   31   75  105  118  126  133  175  179  206  219  225  229  239  245  254   265  267  270  305  312  338  365  428  470  486  554  566  567  619  654   301
          11   11   12   12   12   12   12   12   12   12   12   12   12   13   13    13   13   13   13   13   13   13   13   14   14   14   14   14   14   14   12.7

Philip   14   57   83   89  106  136  155  169  181  188  212  213  221  236  243   247  255  263  306  411  415  431  435  441  522  626  634  642  644  670   308
          10   11   12   12   12   12   12   12   12   12   12   12   12   12   12    12   13   13   13   13   13   13   13   13   14   14   14   14   14   14   12.7

Zalava   42   72  107  163  178  201  202  253  273  292  293  295  297  309  320   326  344  358  368  382  391  400  412  426  432  473  479  483  611  629   327
          11   11   12   12   12   12   12   13   13   13   13   13   13   13   13    13   13   13   13   13   13   13   13   13   13   14   14   14   14   14   12.8

Peru    205  215  226  235  266  269  337  347  353  355  362  372  374  423  446   457  461  463  469  482  591  598  612  623  651  685  687  723  729  767   459
          12   12   12   12   13   13   13   13   13   13   13   13   13   13   13    13   13   13   13   14   14   14   14   14   14   14   14   15   15   15   13.4

Atayal   84  101  124  174  217  241  294  323  340  354  404  407  410  450  465   506  527  530  605  610  636  643  657  671  684  695  739  747  768    .    462
          12   12   12   12   12   12   13   13   13   13   13   13   13   13   13    14   14   14   14   14   14   14   14   14   14   14   15   15   15        13.6

Bushma   52  122  159  182  193  244  291  335  387  417  425  434  449  452  475   484  492  508  528  543  544  560  613  641  683  694  726  728  759  800   464
          11   12   12   12   13   13   13   13   13   13   13   13   13   13   13    14   14   14   14   14   14   14   14   14   14   15   15   15   15   15   13.5

Norse   145  147  176  209  233  264  272  283  298  300  333  370  380  460  481   499  538  548  577  608  614  638  639  701  761  771  773  797  798  817   481
          12   12   12   12   12   13   13   13   13   13   13   13   13   13   14    14   14   14   14   14   14   14   14   15   15   15   15   15   15   16   13.6

SJapan  123  164  234  242  327  332  345  350  356  361  366  421  427  471  480   523  550  569  571  575  578  581  645  649  712  717  718  721  738  839   490
          12   12   12   12   13   13   13   13   13   13   13   13   13   13   14    14   14   14   14   14   14   14   14   15   15   15   15   15   15   17   13.6

Hawaii  153  250  257  261  282  289  311  319  399  401  406  490  502  504  511   558  559  562  576  593  665  668  698  707  708  709  722  742  779  813   513
          12   13   13   13   13   13   13   13   13   13   13   14   14   14   14    14   14   14   14   14   14   15   15   15   15   15   15   15   16   13.7

Hainan  171  195  287  299  346  349  352  398  458  462  512  515  516  524  537   540  542  596  609  618  631  666  672  699  743  777  786  799  807  815   541
          12   12   13   13   13   13   13   13   13   14   14   14   14   14   14    14   14   14   14   14   14   14   15   15   15   15   15   16   16   13.9

Teita   135  165  218  240  341  373  376  420  455  491  534  536  547  556  579   622  628  652  658  675  691  713  716  736  770  778  793  801  821  837   562
          12   12   12   12   13   13   13   13   14   14   14   14   14   14   14    14   14   14   14   14   15   15   15   15   15   15   15   16   17   14.0

Anyang   98  128  222  379  381  392  448  509  510  531  545  552  563  565  637   656  664  667  673  679  693  700  719  732  751  769  783  789  818  834   580
          12   12   12   13   13   13   13   14   14   14   14   14   14   14   14    14   14   14   14   14   15   15   15   15   15   15   15   16   17   14.1

Guam    246  307  408  413  419  433  440  447  494  496  505  529  549  553  572   586  601  604  625  627  659  744  745  760  765  766  774  781  796  802   582
          12   13   13   13   13   13   13   13   14   14   14   14   14   14   14    14   14   14   14   14   15   15   15   15   15   15   15   15   14.0

Dogon   186  232  277  285  438  476  478  520  525  526  532  555  573  592  597   620  621  630  661  678  689  706  714  753  776  792  804  814  816  833   590
          12   12   13   13   13   13   13   14   14   14   14   14   14   14   14    14   14   14   14   14   15   15   15   15   16   16   16   16   14.1

Egypt   258  329  330  386  393  394  444  466  489  500  535  561  607  617  669   682  686  688  696  703  705  715  731  741  746  748  756  762  780  784   597
          13   13   13   13   13   13   13   13   14   14   14   14   14   14   14    14   14   14   15   15   15   15   15   15   15   15   15   15   15   14.1

Easter   61  198  314  322  359  397  451  456  539  564  585  594  602  606  650   653  660  692  724  725  730  737  740  764  782  788  790  791  810  819   597
          11   12   13   13   13   13   13   14   14   14   14   14   14   14   14    14   14   15   15   15   15   15   15   15   15   15   15   16   16   14.1

Buriat  116  203  364  396  439  454  472  495  503  514  582  590  595  600  616   648  663  710  711  720  733  734  735  754  755  758  787  795  803  820   602
          12   12   13   13   13   13   13   14   14   14   14   14   14   14   14    14   14   15   15   15   15   15   15   15   15   15   16   16   16   14.1

Andama  194  325  369  371  383  513  551  584  589  615  635  640  646  647  655   662  676  677  681  690  697  727  749  757  763  808  824  827  828  829   630
          12   13   13   13   13   14   14   14   14   14   14   14   14   14   14    14   14   14   14   15   15   15   15   15   16   16   16   16   14.4

Eskimo  405  488  518  557  574  632  674  680  704  750  752  772  775  785  794   805  806  809  811  812  822  823  825  826  830  831  832  835  836  838   740
          13   14   14   14   14   14   14   14   15   15   15   15   15   15   15    16   16   16   16   16   16   16   16   16   16   16   17   17   17   15.3
```

TABLE 9C-18
POPKIN results: La Ferrassie #2903

```
S.Cruz    7   10   17   21   29   32   35   38   40   46   60   64   72   74   76     97   98  102  105  106  109  111  121  141  157  167  206  223  256  366   100
         7.2  7.4  7.7  7.7  7.9  8.0  8.0  8.2  8.2  8.3  8.5  8.5  8.6  8.6  8.6    8.9  8.9  8.9  8.9  9.0  9.0  9.0  9.1  9.2  9.3  9.4  9.6  9.7  9.8   10   8.7

Arikar    9   23   43   45   49   50   55   65   71   73   78   83   99  119  123    126  129  143  156  159  178  224  226  251  265  290  332  353  390  567   159
         7.4  7.8  8.3  8.3  8.4  8.4  8.5  8.6  8.6  8.6  8.6  8.7  8.9  9.1  9.1    9.1  9.1  9.2  9.3  9.3  9.4  9.7  9.7  9.8  9.9   10   10   10   10   11   9.1

Austra    5    6    8   14   31   36   57   84   90   94  103  115  117  124  146    161  168  200  216  219  233  235  238  302  316  317  318  320  374  426   169
         7.2  7.2  7.3  7.6  7.9  8.1  8.5  8.7  8.8  8.8  8.9  9.0  9.0  9.1  9.3    9.3  9.4  9.5  9.6  9.7  9.7  9.7  9.7   10   10   10   10   10   10   11   9.1

Morior   15   19   20   34   37   41   56   68   69   85   86  108  130  132  136    140  147  151  192  193  207  222  262  276  295  301  321  356  501  503   170
         7.6  7.7  7.7  8.0  8.1  8.2  8.5  8.6  8.6  8.7  8.7  9.0  9.2  9.2  9.2    9.2  9.3  9.3  9.5  9.5  9.6  9.7  9.8  9.9   10   10   10   10   11   11   9.2

Tasman    2    4   12   13   26   42   47   66   80   81   82   89  120  125  144    165  181  185  204  217  228  243  247  286  309  310  389  407  497  802   187
         6.7  7.2  7.5  7.6  7.8  8.3  8.4  8.6  8.7  8.7  8.7  8.8  9.1  9.1  9.2    9.4  9.4  9.5  9.5  9.6  9.7  9.7  9.8   10   10   10   10   11   11   13   9.2

Berg      1   16   22   25   33   54   67   75   91   95  110  116  133  135  155    158  160  171  187  236  240  253  264  285  323  330  381  457  529  540   188
         6.3  7.7  7.7  7.8  8.0  8.5  8.6  8.6  8.8  8.8  9.0  9.0  9.2  9.2  9.3    9.3  9.3  9.4  9.5  9.7  9.7  9.8  9.9   10   10   10   11   11   11   11   9.2

Zalava    3   28   44   48   62   70   87  112  113  118  127  163  174  194  199    227  244  248  254  255  271  273  293  294  299  328  376  467  492  505   212
         7.1  7.8  8.3  8.4  8.5  8.6  8.7  9.0  9.0  9.1  9.1  9.3  9.4  9.5  9.5    9.7  9.8  9.8  9.8  9.8  9.9  9.9   10   10   10   10   10   11   11   11   9.4

Norse    18   24   51   61   93   96  114  137  139  145  152  164  169  170  173    177  197  249  267  279  314  378  395  440  451  482  489  507  524  633   250
         7.7  7.8  8.4  8.5  8.8  8.9  9.0  9.2  9.2  9.2  9.3  9.3  9.4  9.4  9.4    9.4  9.5  9.8  9.9  9.9   10   10   11   11   11   11   11   11   11   12   9.7

Ainu     53  100  107  131  150  175  183  184  190  195  231  241  277  281  288    304  326  340  358  404  414  421  422  499  553  579  607  665  720  814   340
         8.5  8.9  9.0  9.2  9.3  9.4  9.5  9.5  9.5  9.5  9.7  9.7  9.9   10   10     10   10   10   10   11   11   11   11   11   11   11   11   12   12   13  10.2

NJapan   27  101  104  172  179  188  213  214  221  269  270  297  308  345  354    424  441  445  477  487  500  515  522  541  566  571  576  591  608  692   374
         7.8  8.9  8.9  9.4  9.4  9.5  9.6  9.6  9.7  9.9  9.9   10   10   10   10     11   11   11   11   11   11   11   11   11   11   11   11   11   11   12  10.3

Zulu     52   58   59  142  176  180  229  234  250  278  280  312  313  335  361    363  385  430  449  470  471  490  496  519  592  623  675  711  734  744   380
         8.4  8.5  8.5  9.2  9.4  9.4  9.7  9.7  9.8  9.9   10   10   10   10   10     10   10   11   11   11   11   11   11   11   11   12   12   12   12  12.10.4

Tolai    77   88   92  205  209  218  260  283  284  306  319  324  329  338  371    408  411  416  452  466  473  474  479  491  549  575  580  602  637  778   383
         8.6  8.8  8.8  9.6  9.6  9.6  9.8   10   10   10   10   10   10   10   10     11   11   11   11   11   11   11   11   11   11   11   11   12   12  10.4

Peru    148  166  203  212  246  257  263  272  300  307  322  325  333  346  351    383  397  406  412  418  458  510  516  523  557  581  593  618  620  688   391
         9.3  9.4  9.5  9.6  9.8  9.8  9.8  9.9   10   10   10   10   10   10   10     10   11   11   11   11   11   11   11   11   11   11   11   11   11   12  10.5

Philip   11  138  153  162  198  201  215  258  287  296  362  370  372  400  401    410  417  419  420  461  463  478  483  506  551  596  651  661  707  708   392
         7.5  9.2  9.3  9.3  9.5  9.5  9.6  9.8   10   10   10   10   10   10   11     11   11   11   11   11   11   11   11   11   11   11   12   12   12   12  10.4

Egypt   122  128  154  189  191  202  211  282  337  342  367  369  384  387  399    409  413  469  476  488  514  545  546  559  598  611  630  662  666  746   413
         9.1  9.1  9.3  9.5  9.5  9.5  9.6   10   10   10   10   10   10   10   11     11   11   11   11   11   11   11   11   11   11   11   12   12   12  10.6

Buriat   63   79  186  230  239  245  261  275  311  341  348  349  350  375  382    393  423  456  465  475  560  563  569  578  588  589  646  697  699  713   415
         8.5  8.6  9.5  9.7  9.7  9.8  9.8  9.9   10   10   10   10   10   10   10     11   11   11   11   11   11   11   11   11   11   11   11   12   12  12.10.6

Bushma  149  182  220  232  237  336  355  368  459  460  493  521  525  536  542    577  584  597  600  601  627  639  641  645  724  729  737  793  798  831   525
         9.3  9.5  9.7  9.7  9.7   10   10   10   11   11   11   11   11   11   11     11   11   11   11   12   12   12   12   12   12   12   13   13   14  11.1

Hawaii  208  225  292  343  347  373  377  394  396  447  468  480  494  495  498    508  562  587  599  644  655  673  677  690  698  733  747  754  766  790   531
         9.6  9.7   10   10   10   10   10   11   11   11   11   11   11   11   11     11   11   11   12   12   12   12   12   12   12   12   12   12   13  11.1

Teita   196  242  259  266  331  334  391  428  429  442  446  454  484  531  550    606  612  616  626  649  652  659  680  687  691  702  726  728  796  821   534
         9.5  9.7  9.8  9.9   10   10   11   11   11   11   11   11   11   11   11     11   11   11   12   12   12   12   12   12   12   12   12   13   13  11.2

Guam    298  315  352  364  379  388  427  431  432  437  448  450  462  486  526    528  594  604  628  632  671  694  705  727  745  748  755  759  774  783   551
          10   10   10   10   10   10   11   11   11   11   11   11   11   11   11     11   11   11   12   12   12   12   12   12   12   12   12   12   13  11.2

SJapan  252  291  360  380  392  436  455  472  481  485  509  517  544  558  595    615  636  663  667  669  670  682  684  719  739  760  765  768  787  839   580
         9.8   10   10   11   11   11   11   11   11   11   11   11   11   11   11     12   12   12   12   12   12   12   12   12   12   12   13   13   15  11.4

Hainan  210  305  386  402  415  433  511  512  518  538  543  555  583  585  619    622  624  642  654  658  674  695  701  704  763  767  770  780  795  823   619
         9.6   10   10   11   11   11   11   11   11   11   11   11   11   11   11     12   12   12   12   12   12   12   12   12   12   12   13   13  11.5

Andama  134  274  365  425  439  453  530  534  537  552  556  572  609  610  614    617  621  650  672  717  732  735  740  750  756  784  788  806  824  836   608
         9.2  9.9   10   11   11   11   11   11   11   11   11   11   11   11   11     11   11   12   12   12   12   12   12   12   13   13   13   13   14  11.6

Atayal  268  359  403  435  438  443  527  532  539  547  574  590  634  648  653    676  683  685  696  706  716  736  741  753  761  771  782  799  828    .   618
         9.9   10   11   11   11   11   11   11   11   11   11   11   12   12   12     12   12   12   12   12   12   12   12   12   12   13   13   13    .  11.6

Anyang   30   39  289  504  520  533  565  573  582  613  625  629  638  647  681    693  700  703  715  723  725  730  731  738  786  797  804  805  810  832   625
         7.9  8.2   10   11   11   11   11   11   11   12   12   12   12   12   12     12   12   12   12   12   12   12   12   13   13   13   13   13   14  11.6

Eskimo  327  339  398  405  434  444  502  513  554  605  668  679  689  709  712    742  757  762  779  781  785  789  794  800  812  827  829  830  833  835   664
          10   10   11   11   11   11   11   11   11   12   12   12   12   12   12     12   12   12   12   13   13   13   13   13   14   14   14   14  12.0

Easter  303  344  464  535  548  564  568  570  603  656  657  710  714  721  722    749  752  772  773  791  792  803  808  809  811  815  817  820  825  834   688
          10   10   11   11   11   11   11   11   11   12   12   12   12   12   12     12   12   13   13   13   13   13   13   13   13   13   13   13   14  12.1

Dogon   357  561  586  631  635  640  643  660  664  678  686  718  743  751  758    764  769  775  776  777  801  807  813  816  818  819  822  826  837  838   726
          10   11   11   12   12   12   12   12   12   12   12   12   12   12   12     12   12   12   12   12   13   13   13   13   13   13   13   14   14  12.4
```

TABLE 9C-19

POPKIN results: Shanidar #1778

```
Arikar   2    5    8    9   19   20   22   23   38   46   47   48   51   56   73    92   95  109  110  127  152  168  180  194  196  251  269  272  284  457   114
        7.0  7.4  7.7  7.7  8.1  8.1  8.1  8.1  8.5  8.6  8.6  8.6  8.6  8.7  8.9   9.0  9.0  9.1  9.1  9.2  9.3  9.4  9.5  9.6  9.6  9.9   10   10   10   11   8.9

Morior   3   15   25   27   28   29   52   54   60   62   63   66   76  108  123   139  158  160  162  163  175  211  225  237  245  249  305  438  439  499   153
        7.3  8.0  8.2  8.2  8.3  8.3  8.7  8.7  8.7  8.7  8.7  8.8  8.9  9.1  9.2   9.3  9.4  9.4  9.4  9.4  9.5  9.7  9.8  9.8  9.9  9.9   10   11   11   11   9.2

Berg     1   10   11   16   21   39   58   68   69   80   82   85   86  114  119   131  133  134  156  176  184  205  255  264  274  314  395  451  462  472   162
        6.4  7.8  7.8  8.0  8.1  8.5  8.7  8.8  8.8  8.9  8.9  9.0  9.0  9.1  9.2   9.2  9.2  9.2  9.4  9.5  9.6  9.7  9.9   10   10   10   11   11   11   11   9.2

S.Cruz  32   33   34   36   37   49   71   87   99  100  118  135  144  145  153   154  191  197  208  256  266  268  273  286  300  320  361  375  391  400   184
        8.4  8.4  8.4  8.4  8.5  8.6  8.9  9.0  9.1  9.1  9.2  9.2  9.3  9.3  9.3   9.4  9.6  9.6  9.7  9.9   10   10   10   10   10   10   10   10   11   11   9.5

Zalava  12   14   17   18   41   70   89   98  107  120  129  137  141  183  201   207  230  233  244  258  262  302  312  319  335  344  367  386  459  484   204
        7.9  8.0  8.0  8.0  8.5  8.8  9.0  9.0  9.1  9.2  9.2  9.3  9.3  9.6  9.7   9.7  9.8  9.8  9.9  9.9   10   10   10   10   10   10   10   11   11   11   9.5

Tasman   7   13   30   42   50   53   74   81   83   88   93   94  104  105  106   121  125  128  192  231  248  278  381  408  411  432  437  449  458  808   204
        7.7  7.9  8.3  8.5  8.6  8.7  8.9  8.9  8.9  9.0  9.0  9.1  9.1  9.1  9.1   9.2  9.2  9.2  9.6  9.8  9.9   10   10   11   11   11   11   11   11   13   9.5

Norse   24   31   61   75   77   90  116  140  143  151  167  178  224  242  265   283  288  291  304  307  357  384  396  420  485  544  551  599  652  669   284
        8.1  8.3  8.7  8.9  8.9  9.0  9.1  9.3  9.3  9.3  9.4  9.5  9.8  9.8   10    10   10   10   10   11   11   11   11   11   11   11   11   11   12   12  10.0

NJapan   4   40   55   59   97  113  161  169  173  204  215  241  257  260  280   289  296  299  329  334  373  414  440  494  501  520  535  557  578  660   298
        7.3  8.5  8.7  8.7  9.0  9.1  9.4  9.5  9.5  9.7  9.7  9.8  9.9   10   10    10   10   10   10   10   11   11   11   11   11   11   11   11   12  10.0

Peru    57  101  122  124  149  179  187  212  223  228  247  253  271  293  317   324  332  341  372  387  393  398  401  418  442  446  476  492  597  633   314
        8.7  9.1  9.2  9.2  9.3  9.5  9.6  9.7  9.8  9.8  9.9  9.9   10   10   10    10   10   10   10   11   11   11   11   11   11   11   11   11   11   12  10.2

Philip   6   67   96  102  117  148  166  174  186  227  252  263  276  308  328   352  368  374  385  427  448  454  470  491  516  530  575  615  683  745   338
        7.5  8.8  9.0  9.1  9.2  9.3  9.4  9.5  9.6  9.8  9.9   10   10   10   10    10   10   10   11   11   11   11   11   11   11   11   12   12   12  10.2

Zulu    26   45   91  130  132  142  146  165  198  202  277  301  322  331  333   337  345  363  402  466  480  545  546  590  601  608  618  626  715  718   360
        8.2  8.6  9.0  9.2  9.2  9.3  9.3  9.4  9.6  9.7   10   10   10   10   10    10   10   10   11   11   11   11   11   11   12   12   12   12   12   12  10.3

Austra  43   64   79  147  164  193  220  254  279  282  297  318  338  351  364   371  394  441  496  505  515  567  577  585  589  613  636  639  703  778   393
        8.5  8.8  8.9  9.3  9.4  9.6  9.7  9.9   10   10   10   10   10   10   10    10   11   11   11   11   11   11   11   11   11   12   12   12   12   13  10.5

Tolai   72   78  115  126  219  235  259  292  310  311  321  327  343  355  369   421  423  455  477  508  539  556  564  571  581  583  610  629  694  789   404
        8.9  8.9  9.1  9.2  9.7  9.8  9.9   10   10   10   10   10   10   10   10    11   11   11   11   11   11   11   11   11   11   12   12   12   13  10.6

Ainu    44  107  101  105  214  229  276  290  296  366  367  603  616  617  625   634  636  643  681  693  697  541  542  554  545  648  729  741  763  794   429
        8.6  9.1  9.5  9.6  9.7  9.8  9.8   10   10   10   10   11   11   11   11    11   11   11   11   11   11   11   11   11   11   12   12   12   12   13  10.7

Hawaii  84  112  150  159  199  232  240  243  281  323  340  376  383  406  419   503  532  533  555  572  573  584  607  643  664  666  685  700  714  768   442
        9.0  9.1  9.3  9.4  9.6  9.8  9.8  9.8   10   10   10   11   11   11   11    11   11   11   11   11   11   11   12   12   12   12   12   12   12   12  10.7

Egypt  155  170  171  190  213  218  226  298  316  360  426  431  447  465  483   490  511  561  580  614  622  637  640  644  650  672  695  705  722  731   468
        9.4  9.5  9.5  9.6  9.7  9.7  9.8   10   10   11   11   11   11   11   11    11   11   11   12   12   12   12   12   12   12   12   12   12   12   12  10.9

SJapan 157  216  270  295  313  315  330  336  353  365  370  378  382  422  482   489  531  536  538  549  553  558  570  619  624  667  670  716  721  836   472
        9.4  9.7   10   10   10   10   10   10   10   10   10   10   11   11   11    11   11   11   11   11   11   11   12   12   12   12   12   12   12   15  11.0

Buriat 138  172  177  185  203  234  267  415  433  445  471  474  478  486  504   507  514  534  537  543  579  594  605  653  674  677  679  696  726  770   479
        9.3  9.5  9.5  9.6  9.7  9.8   10   11   11   11   11   11   11   11   11    11   11   11   11   11   11   11   12   12   12   12   12   12   12   12  10.9

Hainan 136  209  222  325  326  358  362  379  389  407  444  452  467  468  475   479  500  506  569  574  596  645  657  659  701  709  725  751  753  780   501
        9.3  9.7  9.7   10   10   10   10   10   11   11   11   11   11   11   11    11   11   11   11   11   11   12   12   12   12   12   12   12   12   13  11.0

Atayal 188  200  285  348  350  392  399  404  405  424  429  463  512  521  523   528  565  576  582  600  612  628  630  631  656  662  671  750  805    .   508
        9.6  9.6   10   10   10   11   11   11   11   11   11   11   11   11   11    11   11   11   11   11   12   12   12   12   12   12   12   12   13    .  11.1

Easter 111  182  217  246  342  390  410  412  430  487  488  509  518  525  529   552  598  604  606  616  647  654  668  687  717  736  747  774  796  814   534
        9.1  9.5  9.7  9.9   10   11   11   11   11   11   11   11   11   11   11    11   11   11   12   12   12   12   12   12   12   12   13   13   13   13  11.2

Anyang  35   65  238  250  303  397  456  460  473  502  526  540  547  550  559   586  588  595  646  649  658  675  676  693  723  749  762  765  786  817   534
        8.4  8.8  9.8  9.9   10   11   11   11   11   11   11   11   11   11   11    11   11   11   12   12   12   12   12   12   12   12   13   13   13   13  11.2

Teita  189  210  275  339  349  354  377  428  435  461  498  513  519  638  642   684  686  688  689  697  719  735  740  743  755  764  783  798  802  831   578
        9.6  9.7   10   10   10   10   10   11   11   11   11   11   12   12   12    12   12   12   12   12   12   12   12   12   13   13   13   13   14  11.5

Guam   239  287  306  366  413  464  510  517  522  548  563  566  591  609  621   625  632  648  673  702  706  707  711  744  746  757  759  785  787  804   597
        9.8   10   10   10   11   11   11   11   11   11   11   11   12   12   12    12   12   12   12   12   12   12   12   12   12   13   13   13   13  11.5

Bushma 206  261  356  359  409  453  560  587  635  655  678  690  699  704  710   713  720  727  733  737  766  772  775  781  784  788  800  813  819  837   651
        9.7   10   10   10   11   11   11   11   12   12   12   12   12   12   12    12   12   12   12   12   13   13   13   13   13   13   13   13   15  11.9

Andama 221  309  380  450  469  527  602  603  617  620  623  641  651  663  680   681  691  712  724  742  758  767  769  773  790  792  803  821  824  830   651
        9.7   10   10   11   11   11   12   12   12   12   12   12   12   12   12    12   12   12   13   13   13   13   13   13   13   14   14   14   14  11.9

Dogon  388  524  562  592  665  692  708  734  738  739  752  754  760  761  771   791  793  799  801  806  809  810  811  818  826  827  829  832  838  839   746
         11   11   11   11   12   12   12   12   12   12   12   12   12   12   12    13   13   13   13   13   13   13   13   13   14   14   14   14   15  12.7

Eskimo 495  568  611  627  634  661  682  728  730  732  748  756  776  777  779   782  795  797  807  812  815  816  820  822  823  825  828  833  834  835   752
         11   11   12   12   12   12   12   12   12   12   12   12   13   13   13    13   13   13   13   13   13   13   13   14   14   14   14   14   14  12.7
```

POPKIN results: Broken Hill #1712

```
Austra   3   4   5   7   9  11  12  16  18  21  22  23  27  30  31    33  38  41  43  46  48  57  61  77  78  85  97 109 171 176    47
        7.5 7.5 7.6 8.0 8.1 8.2 8.4 8.7 8.7 8.8 8.8 8.8 8.9 9.0 9.0   9.1 9.2 9.3 9.4 9.4 9.4 9.6 9.7 9.9 9.9  10  10  10  11  11   9.1

S.Cruz  10  17  19  20  26  32  34  36  39  44  50  55  62  64  70    72  74  75  81  88  99 108 111 132 156 163 190 237 262 311    91
        8.2 8.7 8.7 8.8 8.9 9.1 9.1 9.2 9.3 9.4 9.4 9.5 9.7 9.7 9.9   9.9 9.9 9.9 9.9 9.9  10  10  10  10  10  10  11  11  11  11   9.8

Tasman   1   2   6   8  24  25  28  37  49  56  67  69  76 102 113   115 120 123 128 131 134 152 166 169 184 204 226 399 544 623   136
        7.2 7.4 7.9 8.0 8.9 8.9 9.0 9.2 9.4 9.6 9.8 9.8 9.9  10  10    10  10  10  10  10  10  10  11  11  11  11  11  12  12  13   9.9

Tolai   66  71  79  86  94 107 119 122 136 148 150 155 165 174 181   221 292 305 306 315 322 335 346 359 375 394 405 414 458 475   239
        9.8 9.9 9.9  10  10  10  10  10  10  10  10  10  11  11  11    11  11  11  11  11  11  11  12  12  12  12  12  12  12  12  10.9

Morior  14  42  47  59  82  95 106 110 125 142 149 151 182 246 270   282 293 296 317 328 356 376 377 389 438 443 489 562 563 587   264
        8.5 9.3 9.4 9.7 9.9  10  10  10  10  10  10  10  11  11  11    11  11  11  11  11  12  12  12  12  12  12  12  13  13  13  10.9

Zalava  13  84 101 104 137 138 147 162 179 180 183 195 199 219 229   238 239 274 275 316 329 334 348 353 396 397 478 496 505 748   267
        8.4  10  10  10  10  10  11  11  11  11  11  11  11  11  11    11  11  11  11  11  11  12  12  12  12  12  12  12  13  11.0

Arikar  63  92 112 114 139 160 161 185 207 213 227 236 240 257 271   288 290 291 297 304 323 380 402 419 426 430 452 482 662 708   294
        9.7  10  10  10  10  11  11  11  11  11  11  11  11  11  11    11  11  11  11  11  11  12  12  12  12  12  12  12  13  13  11.2

Peru    68 103 105 129 144 178 200 210 212 215 222 224 235 247 251   255 266 281 309 326 327 365 423 425 450 469 480 512 614 655   297
        9.8  10  10  10  10  11  11  11  11  11  11  11  11  11  11    11  11  11  11  11  11  12  12  12  12  12  12  12  13  13  11.2

Bushma  54  93 100 140 141 186 191 216 218 231 234 243 254 285 301   314 320 337 347 351 354 355 358 362 420 460 504 517 558 785   308
        9.5  10  10  10  10  11  11  11  11  11  11  11  11  11  11    11  11  11  12  12  12  12  12  12  12  12  12  12  14  11.3

Berg    29  45  53  90  91 121 133 143 173 177 197 202 208 214 286   294 295 366 373 374 386 400 449 538 561 591 631 664 743 779   320
        9.0 9.4 9.5  10  10  10  10  11  11  11  11  11  11  11  11    11  11  12  12  12  12  12  12  13  13  13  13  13  14  11.3

Zulu    58  60  89  96 116 146 168 175 196 217 253 258 259 260 272   284 324 330 383 385 390 395 417 454 484 596 643 654 724 738   327
        9.7 9.7  10  10  10  10  11  11  11  11  11  11  11  11  11    11  11  11  12  12  12  12  12  12  13  13  13  13  13  11.3

Norse   51  52  87 124 135 158 164 187 192 193 201 203 220 232 248   283 313 360 361 370 379 393 551 553 564 588 638 686 701 713   330
        9.4 9.5  10  10  10  11  11  11  11  11  11  11  11  11  11    11  11  12  12  12  12  12  12  12  13  13  13  13  13  11.3

Ainu    80  98 126 153 157 159 172 206 244 252 279 307 338 340 341   344 401 421 433 437 440 461 466 487 498 627 632 642 650 812   367
        9.9  10  10  10  11  11  11  11  11  11  11  11  11  11  11    12  12  12  12  12  12  12  12  12  12  13  13  13  13  14  11.6

NJapan  15  65  83 117 118 145 198 241 250 269 277 308 352 364 369   406 412 427 455 465 516 522 525 529 541 578 585 593 610 776   370
        8.6 9.8 9.9  10  10  10  11  11  11  11  11  11  12  12  12    12  12  12  12  12  12  12  12  12  12  13  13  13  13  14  11.5

Philip  40 170 223 228 230 233 256 264 276 300 310 312 318 319 331   357 363 384 411 472 483 495 501 513 549 572 575 609 634 684   380
        9.3  11  11  11  11  11  11  11  11  11  11  12  12  12  12    12  12  12  12  12  12  12  12  13  13  13  13  13  13  11.6

Teita  127 130 225 245 249 267 268 333 339 342 391 418 441 448 464   473 506 507 523 557 597 606 608 612 628 669 671 694 790 831   465
         10  10  11  11  11  11  11  11  11  11  12  12  12  12  12    12  12  12  12  12  13  13  13  13  13  13  13  13  14  15  12.1

Egypt  261 287 321 325 336 345 372 409 436 453 494 509 515 530 536   569 577 581 617 624 626 630 639 659 689 700 714 723 761 771   534
         11  11  11  11  11  12  12  12  12  12  12  12  12  12  12    13  13  13  13  13  13  13  13  13  13  13  13  13  14  14  12.4

Atayal 167 188 189 302 349 371 388 404 413 463 479 520 548 554 571   595 600 607 646 656 658 676 683 698 707 722 760 766 818   .   534
         11  11  11  11  12  12  12  12  12  12  12  12  12  12  13    13  13  13  13  13  13  13  13  13  13  13  14  14  14   .  12.4

Guam   273 350 398 407 408 410 432 435 447 459 468 477 486 491 503   514 528 534 539 542 601 616 685 716 720 732 746 747 768 809   541
         11  12  12  12  12  12  12  12  12  12  12  12  12  12  12    12  12  12  12  13  13  13  13  13  13  13  13  13  14  14  12.4

Anyang  35  73 265 381 434 451 476 481 493 511 521 550 568 589 604   611 613 618 621 625 673 679 692 711 715 726 753 765 813 815   562
        9.2 9.9  11  12  12  12  12  12  12  12  12  12  13  13  13    13  13  13  13  13  13  13  13  13  13  14  14  14  14  14  12.5

Hainan 278 298 299 343 387 457 485 502 508 510 532 552 565 573 580   598 622 636 670 681 709 718 725 733 734 759 773 783 788 805   587
         11  11  11  12  12  12  12  12  12  12  12  12  13  13  13    13  13  13  13  13  13  13  13  13  13  14  14  14  14  14  12.7

Hawaii 205 289 367 392 429 439 442 445 470 499 527 543 566 586 599   635 644 648 657 712 717 727 728 737 754 755 777 786 801 824   590
         11  11  12  12  12  12  12  12  12  12  12  12  13  13  13    13  13  13  13  13  13  13  13  13  14  14  14  14  14  15  12.7

SJapan 242 263 378 403 462 474 490 497 518 546 555 556 567 576 583   594 640 672 706 739 741 744 745 751 758 763 769 791 823 838   606
         11  11  12  12  12  12  12  12  12  12  12  12  13  13  13    13  13  13  13  13  13  13  13  14  14  14  14  14  15  16  12.9

Andama 154 209 332 382 416 444 492 526 535 540 582 603 620 652 661   663 674 677 678 687 695 703 730 740 742 762 814 822 834 836   607
         10  11  11  12  12  12  12  12  12  12  13  13  13  13  13    13  13  13  13  13  13  13  13  13  14  14  14  14  15  15  12.9

Buriat 194 211 415 422 471 545 559 574 579 629 637 666 693 696 697   721 735 750 764 767 772 780 789 793 794 797 810 811 816 825   657
         11  11  12  12  12  12  12  12  13  13  13  13  13  13  13    13  13  13  14  14  14  14  14  14  14  14  14  14  14  15  13.1

Easter 280 303 428 431 467 519 533 547 619 649 651 653 665 688 690   691 710 749 752 757 778 781 787 796 803 807 817 827 828 832   661
         11  11  12  12  12  12  12  13  13  13  13  13  13  13  13    13  13  14  14  14  14  14  14  14  14  14  15  15  15  13.2

Dogon  424 446 488 524 537 560 584 590 602 633 641 645 660 667 675   680 682 702 704 705 719 736 774 792 795 798 804 820 821 833   668
         12  12  12  12  12  12  13  13  13  13  13  13  13  13  13    13  13  13  13  13  13  14  14  14  14  14  14  14  15  13.2

Eskimo 368 456 500 531 570 592 605 615 647 668 699 729 731 756 770   775 782 784 799 800 802 806 808 819 826 829 830 835 837 839   714
         12  12  12  12  13  13  13  13  13  13  13  13  13  14  14    14  14  14  14  14  14  14  14  14  15  15  15  15  15  16  13.7
```

TABLE 9C-21

POPKIN results: Upper Cave 101 #2718

Arikar	1	20	22	29	30	37	39	43	49	54	60	64	81	105	108	109	126	178	182	226	227	228	236	249	255	259	269	271	362	387	144
	5.5	6.4	6.4	6.5	6.5	6.6	6.7	6.7	6.8	6.8	6.8	6.9	7.0	7.2	7.2	7.2	7.3	7.6	7.6	7.8	7.8	7.8	7.8	7.9	7.9	7.9	8.0	8.0	8.4	8.4	7.2

| Berg | 12 | 18 | 27 | 33 | 51 | 58 | 65 | 69 | 74 | 76 | 79 | 87 | 91 | 101 | 106 | 124 | 133 | 145 | 147 | 192 | 230 | 247 | 254 | 285 | 290 | 328 | 342 | 351 | 379 | 625 | 167 |
| | 6.2 | 6.4 | 6.5 | 6.6 | 6.8 | 6.8 | 6.9 | 7.0 | 7.0 | 7.0 | 7.0 | 7.1 | 7.1 | 7.2 | 7.2 | 7.3 | 7.3 | 7.4 | 7.4 | 7.7 | 7.8 | 7.9 | 7.9 | 8.0 | 8.1 | 8.3 | 8.3 | 8.3 | 8.4 | 9.4 | 7.4 |

| Easter | 6 | 23 | 36 | 47 | 48 | 50 | 53 | 63 | 78 | 82 | 84 | 86 | 88 | 95 | 123 | 142 | 153 | 160 | 162 | 198 | 209 | 222 | 244 | 263 | 308 | 310 | 321 | 343 | 504 | 520 | 167 |
| | 5.9 | 6.4 | 6.6 | 6.8 | 6.8 | 6.8 | 6.8 | 6.9 | 7.0 | 7.1 | 7.1 | 7.1 | 7.1 | 7.1 | 7.3 | 7.4 | 7.4 | 7.5 | 7.5 | 7.7 | 7.7 | 7.8 | 7.9 | 7.9 | 8.2 | 8.2 | 8.2 | 8.3 | 8.9 | 8.9 | 7.4 |

| Zalava | 3 | 5 | 7 | 11 | 15 | 21 | 26 | 38 | 40 | 46 | 80 | 92 | 93 | 113 | 132 | 141 | 143 | 179 | 185 | 212 | 220 | 241 | 280 | 283 | 299 | 304 | 347 | 378 | 521 | 700 | 172 |
| | 5.8 | 5.9 | 6.0 | 6.2 | 6.3 | 6.4 | 6.5 | 6.7 | 6.7 | 6.7 | 7.0 | 7.1 | 7.1 | 7.2 | 7.3 | 7.4 | 7.4 | 7.6 | 7.6 | 7.7 | 7.8 | 7.8 | 8.0 | 8.0 | 8.1 | 8.1 | 8.3 | 8.4 | 8.9 | 9.8 | 7.3 |

| Norse | 2 | 8 | 9 | 10 | 16 | 24 | 28 | 32 | 72 | 73 | 99 | 118 | 122 | 159 | 161 | 167 | 171 | 189 | 197 | 261 | 277 | 284 | 359 | 383 | 388 | 407 | 422 | 469 | 479 | 779 | 206 |
| | 5.6 | 6.0 | 6.1 | 6.1 | 6.3 | 6.5 | 6.5 | 6.6 | 7.0 | 7.0 | 7.1 | 7.3 | 7.3 | 7.4 | 7.5 | 7.5 | 7.5 | 7.6 | 7.7 | 7.9 | 8.0 | 8.0 | 8.3 | 8.4 | 8.4 | 8.5 | 8.6 | 8.8 | 8.8 | 11 | 7.5 |

| Tolai | 17 | 31 | 41 | 56 | 66 | 103 | 104 | 114 | 175 | 205 | 208 | 229 | 237 | 240 | 242 | 266 | 289 | 295 | 302 | 312 | 317 | 344 | 421 | 429 | 445 | 622 | 635 | 666 | 687 | 756 | 295 |
| | 6.3 | 6.6 | 6.7 | 6.8 | 6.9 | 7.2 | 7.2 | 7.2 | 7.5 | 7.7 | 7.7 | 7.8 | 7.8 | 7.8 | 7.8 | 7.9 | 8.1 | 8.1 | 8.1 | 8.2 | 8.2 | 8.3 | 8.6 | 8.6 | 8.7 | 9.4 | 9.4 | 9.6 | 9.7 | 10 | 8.0 |

| Tasman | 42 | 83 | 94 | 96 | 100 | 110 | 112 | 139 | 146 | 148 | 176 | 181 | 186 | 199 | 217 | 234 | 300 | 311 | 346 | 377 | 381 | 408 | 443 | 455 | 491 | 522 | 527 | 611 | 673 | 751 | 295 |
| | 6.7 | 7.1 | 7.1 | 7.1 | 7.1 | 7.2 | 7.2 | 7.3 | 7.4 | 7.4 | 7.6 | 7.6 | 7.6 | 7.7 | 7.7 | 7.8 | 8.1 | 8.2 | 8.3 | 8.4 | 8.4 | 8.5 | 8.7 | 8.7 | 8.8 | 8.9 | 8.9 | 9.3 | 9.6 | 10 | 8.0 |

| Morior | 35 | 57 | 59 | 62 | 115 | 149 | 158 | 188 | 193 | 210 | 248 | 267 | 287 | 292 | 327 | 338 | 355 | 385 | 386 | 430 | 439 | 470 | 483 | 499 | 525 | 619 | 630 | 651 | 696 | 766 | 344 |
| | 6.6 | 6.8 | 6.8 | 6.9 | 7.3 | 7.4 | 7.4 | 7.6 | 7.7 | 7.7 | 7.9 | 8.0 | 8.1 | 8.1 | 8.3 | 8.3 | 8.3 | 8.4 | 8.4 | 8.6 | 8.7 | 8.8 | 8.8 | 8.9 | 8.9 | 9.3 | 9.4 | 9.5 | 9.7 | 10 | 8.2 |

| Ainu | 25 | 85 | 89 | 102 | 129 | 144 | 154 | 156 | 166 | 180 | 194 | 260 | 303 | 392 | 401 | 411 | 413 | 426 | 449 | 450 | 473 | 541 | 552 | 570 | 584 | 597 | 636 | 722 | 746 | 754 | 370 |
| | 6.5 | 7.1 | 7.1 | 7.2 | 7.3 | 7.4 | 7.4 | 7.4 | 7.5 | 7.6 | 7.7 | 7.9 | 8.1 | 8.5 | 8.5 | 8.5 | 8.5 | 8.6 | 8.7 | 8.7 | 8.8 | 9.0 | 9.0 | 9.1 | 9.2 | 9.2 | 9.4 | 10 | 10 | 10 | 8.3 |

| Buriat | 14 | 61 | 107 | 137 | 172 | 195 | 202 | 233 | 235 | 276 | 279 | 306 | 315 | 322 | 337 | 372 | 374 | 384 | 415 | 425 | 474 | 484 | 536 | 590 | 606 | 664 | 714 | 759 | 777 | 787 | 385 |
| | 6.3 | 6.9 | 7.2 | 7.3 | 7.5 | 7.7 | 7.7 | 7.8 | 7.8 | 8.0 | 8.0 | 8.2 | 8.2 | 8.2 | 8.3 | 8.4 | 8.4 | 8.4 | 8.5 | 8.6 | 8.8 | 8.8 | 8.9 | 9.0 | 9.2 | 9.3 | 9.6 | 9.9 | 10 | 11 | 8.4 |

| Hawaii | 68 | 77 | 134 | 152 | 170 | 187 | 207 | 218 | 224 | 288 | 349 | 363 | 365 | 368 | 399 | 406 | 436 | 456 | 457 | 460 | 486 | 500 | 509 | 532 | 569 | 591 | 595 | 675 | 729 | 749 | 391 |
| | 6.9 | 7.0 | 7.3 | 7.4 | 7.5 | 7.6 | 7.7 | 7.8 | 7.8 | 8.1 | 8.3 | 8.4 | 8.4 | 8.4 | 8.5 | 8.5 | 8.7 | 8.7 | 8.7 | 8.8 | 8.8 | 8.9 | 9.0 | 9.1 | 9.2 | 9.2 | 9.6 | 10 | 10 | 8.4 |

| Guam | 55 | 128 | 174 | 201 | 203 | 211 | 238 | 239 | 246 | 275 | 281 | 309 | 323 | 336 | 380 | 389 | 423 | 433 | 448 | 454 | 463 | 466 | 561 | 564 | 578 | 589 | 652 | 665 | 671 | 768 | 391 |
| | 6.8 | 7.3 | 7.5 | 7.7 | 7.7 | 7.7 | 7.8 | 7.8 | 7.9 | 8.0 | 8.0 | 8.2 | 8.2 | 8.3 | 8.4 | 8.4 | 8.6 | 8.6 | 8.7 | 8.7 | 8.7 | 8.7 | 9.1 | 9.1 | 9.2 | 9.2 | 9.5 | 9.6 | 9.6 | 10 | 8.5 |

| Peru | 125 | 168 | 177 | 183 | 221 | 225 | 256 | 258 | 264 | 265 | 270 | 286 | 333 | 356 | 361 | 382 | 402 | 428 | 431 | 434 | 476 | 524 | 534 | 612 | 613 | 620 | 670 | 689 | 712 | 728 | 400 |
| | 7.3 | 7.5 | 7.6 | 7.6 | 7.8 | 7.8 | 7.9 | 7.9 | 7.9 | 7.9 | 8.0 | 8.1 | 8.3 | 8.3 | 8.4 | 8.4 | 8.5 | 8.6 | 8.6 | 8.6 | 8.8 | 8.9 | 9.0 | 9.3 | 9.3 | 9.4 | 9.6 | 9.7 | 9.9 | 10 | 8.5 |

| Eskimo | 71 | 90 | 116 | 119 | 157 | 206 | 216 | 230 | 253 | 291 | 313 | 357 | 397 | 403 | 409 | 444 | 451 | 471 | 472 | 501 | 523 | 526 | 554 | 560 | 561 | 641 | 660 | 676 | 690 | 790 | 408 |
| | 7.0 | 7.1 | 7.3 | 7.3 | 7.4 | 7.7 | 7.8 | 7.8 | 7.9 | 8.1 | 8.2 | 8.3 | 8.5 | 8.5 | 8.5 | 8.7 | 8.7 | 8.8 | 8.8 | 8.9 | 8.9 | 8.9 | 9.0 | 9.2 | 9.2 | 9.5 | 9.6 | 9.7 | 9.8 | 11 | 8.5 |

| Anyang | 19 | 34 | 67 | 163 | 191 | 245 | 273 | 282 | 324 | 348 | 360 | 400 | 410 | 414 | 493 | 511 | 517 | 528 | 538 | 553 | 566 | 568 | 582 | 615 | 642 | 662 | 663 | 690 | 707 | 711 | 436 |
| | 6.4 | 6.6 | 6.9 | 7.5 | 7.6 | 7.9 | 8.0 | 8.0 | 8.3 | 8.3 | 8.3 | 8.5 | 8.5 | 8.5 | 8.8 | 8.9 | 8.9 | 8.9 | 9.0 | 9.0 | 9.1 | 9.1 | 9.2 | 9.3 | 9.5 | 9.6 | 9.6 | 9.7 | 9.8 | 9.9 | 8.6 |

| Atayal | 13 | 136 | 184 | 252 | 293 | 307 | 314 | 330 | 345 | 369 | 370 | 371 | 398 | 404 | 418 | 440 | 462 | 482 | 495 | 505 | 519 | 539 | 549 | 629 | 645 | 653 | 708 | 757 | 761 | . | 436 |
| | 6.3 | 7.3 | 7.6 | 7.9 | 8.1 | 8.2 | 8.3 | 8.3 | 8.4 | 8.4 | 8.4 | 8.5 | 8.5 | 8.6 | 8.7 | 8.7 | 8.8 | 8.8 | 8.9 | 8.9 | 9.0 | 9.0 | 9.4 | 9.5 | 9.5 | 9.8 | 10 | 10 | . | 8.6 |

| Teita | 44 | 45 | 120 | 131 | 169 | 200 | 257 | 262 | 296 | 329 | 341 | 376 | 458 | 459 | 464 | 468 | 480 | 496 | 542 | 558 | 577 | 600 | 679 | 682 | 692 | 735 | 737 | 739 | 758 | 775 | 449 |
| | 6.7 | 6.7 | 7.3 | 7.3 | 7.5 | 7.7 | 7.9 | 7.9 | 8.1 | 8.3 | 8.3 | 8.4 | 8.7 | 8.7 | 8.7 | 8.8 | 8.8 | 8.9 | 9.0 | 9.1 | 9.2 | 9.3 | 9.7 | 9.7 | 9.7 | 10 | 10 | 10 | 10 | 10 | 8.7 |

| Philip | 70 | 130 | 151 | 165 | 190 | 213 | 274 | 319 | 334 | 335 | 339 | 364 | 438 | 442 | 508 | 512 | 544 | 587 | 608 | 616 | 634 | 643 | 667 | 701 | 704 | 713 | 717 | 720 | 748 | 760 | 472 |
| | 7.0 | 7.3 | 7.4 | 7.5 | 7.6 | 7.7 | 8.0 | 8.2 | 8.3 | 8.3 | 8.3 | 8.4 | 8.7 | 8.7 | 8.9 | 8.9 | 9.0 | 9.2 | 9.3 | 9.3 | 9.4 | 9.5 | 9.6 | 9.8 | 9.8 | 9.9 | 9.9 | 9.9 | 10 | 10 | 8.8 |

| S.Cruz | 75 | 111 | 127 | 196 | 214 | 297 | 305 | 320 | 325 | 391 | 427 | 453 | 489 | 494 | 502 | 510 | 518 | 533 | 609 | 614 | 624 | 628 | 669 | 683 | 685 | 695 | 747 | 791 | 795 | 817 | 488 |
| | 7.0 | 7.2 | 7.3 | 7.7 | 7.7 | 8.1 | 8.1 | 8.2 | 8.3 | 8.5 | 8.6 | 8.7 | 8.8 | 8.8 | 8.9 | 8.9 | 8.9 | 9.0 | 9.3 | 9.3 | 9.4 | 9.4 | 9.6 | 9.7 | 9.7 | 9.7 | 10 | 11 | 11 | 11 | 8.9 |

| NJapan | 135 | 140 | 150 | 215 | 219 | 231 | 354 | 373 | 375 | 390 | 417 | 437 | 441 | 447 | 461 | 467 | 497 | 547 | 555 | 557 | 563 | 680 | 705 | 706 | 730 | 752 | 769 | 785 | 796 | 805 | 490 |
| | 7.3 | 7.3 | 7.4 | 7.8 | 7.8 | 7.8 | 8.3 | 8.4 | 8.4 | 8.5 | 8.5 | 8.7 | 8.7 | 8.7 | 8.7 | 8.8 | 8.8 | 9.0 | 9.0 | 9.1 | 9.1 | 9.7 | 9.8 | 9.8 | 10 | 10 | 10 | 11 | 11 | 11 | 8.9 |

| Austra | 97 | 98 | 121 | 272 | 316 | 331 | 352 | 367 | 405 | 419 | 420 | 487 | 488 | 515 | 516 | 540 | 545 | 565 | 571 | 585 | 588 | 593 | 605 | 626 | 632 | 638 | 674 | 724 | 812 | 814 | 491 |
| | 7.1 | 7.1 | 7.3 | 8.0 | 8.2 | 8.3 | 8.3 | 8.4 | 8.5 | 8.6 | 8.6 | 8.8 | 8.8 | 8.9 | 8.9 | 9.0 | 9.0 | 9.1 | 9.1 | 9.2 | 9.2 | 9.2 | 9.3 | 9.4 | 9.4 | 9.5 | 9.6 | 10 | 11 | 11 | 8.9 |

| Hainan | 4 | 117 | 268 | 278 | 318 | 358 | 393 | 394 | 395 | 416 | 452 | 475 | 506 | 507 | 529 | 556 | 559 | 567 | 592 | 594 | 598 | 599 | 603 | 633 | 650 | 654 | 672 | 677 | 697 | 792 | 495 |
| | 5.8 | 7.3 | 8.0 | 8.0 | 8.2 | 8.3 | 8.5 | 8.5 | 8.5 | 8.5 | 8.7 | 8.8 | 8.9 | 8.9 | 8.9 | 9.0 | 9.1 | 9.1 | 9.2 | 9.2 | 9.2 | 9.3 | 9.3 | 9.4 | 9.5 | 9.5 | 9.6 | 9.7 | 9.7 | 11 | 8.8 |

| SJapan | 164 | 223 | 250 | 294 | 326 | 350 | 353 | 424 | 481 | 485 | 530 | 531 | 535 | 537 | 572 | 586 | 627 | 639 | 648 | 694 | 709 | 710 | 725 | 734 | 753 | 764 | 772 | 803 | 815 | 821 | 562 |
| | 7.5 | 7.8 | 7.9 | 8.1 | 8.3 | 8.3 | 8.3 | 8.6 | 8.8 | 8.8 | 8.8 | 8.9 | 9.0 | 9.0 | 9.0 | 9.1 | 9.2 | 9.4 | 9.5 | 9.5 | 9.7 | 9.8 | 9.9 | 10 | 10 | 10 | 10 | 10 | 11 | 11 | 12 | 9.3 |

| Egypt | 155 | 173 | 298 | 332 | 396 | 446 | 465 | 472 | 498 | 503 | 543 | 548 | 551 | 560 | 562 | 574 | 601 | 623 | 647 | 649 | 657 | 681 | 688 | 699 | 726 | 738 | 740 | 765 | 786 | 804 | 563 |
| | 7.4 | 7.5 | 8.1 | 8.3 | 8.5 | 8.7 | 8.7 | 8.8 | 8.9 | 8.9 | 9.0 | 9.0 | 9.0 | 9.1 | 9.1 | 9.2 | 9.3 | 9.4 | 9.5 | 9.5 | 9.5 | 9.7 | 9.7 | 9.8 | 10 | 10 | 10 | 10 | 11 | 11 | 9.2 |

| Zulu | 52 | 138 | 243 | 251 | 340 | 366 | 432 | 435 | 513 | 514 | 575 | 583 | 604 | 618 | 637 | 644 | 646 | 659 | 660 | 661 | 678 | 684 | 686 | 702 | 716 | 718 | 770 | 788 | 800 | 819 | 564 |
| | 6.8 | 7.3 | 7.9 | 7.9 | 8.3 | 8.4 | 8.6 | 8.7 | 8.9 | 8.9 | 9.2 | 9.2 | 9.3 | 9.3 | 9.4 | 9.5 | 9.5 | 9.6 | 9.6 | 9.6 | 9.7 | 9.7 | 9.7 | 9.8 | 9.9 | 9.9 | 10 | 11 | 11 | 11 | 9.3 |

| Bushma | 204 | 301 | 412 | 546 | 576 | 579 | 596 | 610 | 617 | 621 | 631 | 640 | 656 | 703 | 719 | 727 | 731 | 732 | 741 | 742 | 771 | 773 | 780 | 781 | 782 | 794 | 802 | 813 | 818 | 834 | 668 |
| | 7.7 | 8.1 | 8.5 | 9.0 | 9.2 | 9.2 | 9.2 | 9.3 | 9.3 | 9.4 | 9.4 | 9.5 | 9.5 | 9.8 | 9.9 | 10 | 10 | 10 | 10 | 10 | 10 | 11 | 11 | 11 | 11 | 11 | 11 | 11 | 11 | 12 | 9.9 |

| Andama | 477 | 478 | 550 | 607 | 655 | 658 | 691 | 723 | 733 | 744 | 745 | 750 | 755 | 767 | 774 | 776 | 778 | 783 | 784 | 799 | 801 | 806 | 809 | 810 | 811 | 820 | 823 | 826 | 827 | 828 | 740 |
| | 8.8 | 8.9 | 9.0 | 9.3 | 9.5 | 9.5 | 9.7 | 10 | 10 | 10 | 10 | 10 | 10 | 10 | 10 | 11 | 11 | 11 | 11 | 11 | 11 | 11 | 11 | 11 | 11 | 12 | 12 | 12 | 12 | 12 | 10.4 |

| Dogon | 490 | 573 | 602 | 693 | 715 | 721 | 736 | 743 | 762 | 763 | 789 | 793 | 797 | 798 | 807 | 808 | 816 | 822 | 824 | 825 | 829 | 830 | 831 | 832 | 833 | 835 | 836 | 837 | 838 | 839 | 774 |
| | 8.8 | 9.1 | 9.3 | 9.7 | 9.9 | 9.9 | 10 | 10 | 10 | 10 | 11 | 11 | 11 | 11 | 11 | 11 | 11 | 12 | 12 | 12 | 12 | 12 | 12 | 12 | 12 | 12 | 13 | 13 | 13 | 13 | 11.1 |

TABLE 10

POPKIN results: Liujiang

Zalava 1 3 5 6 13 15 16 19 21 23 26 31 36 41 44   51 83 88 92 96 108 156 167 175 180 183 188 321 363 523   102
2.6 3.1 3.4 3.4 3.8 3.8 3.9 3.9 4.0 4.0 4.1 4.1 4.1 4.2 4.3   4.3 4.8 4.8 4.8 4.9 4.9 5.2 5.3 5.3 5.3 5.4 5.4 5.8 6.0 6.5   4.5

Norse 7 8 9 17 18 22 32 34 46 67 71 105 109 144 181   190 216 221 226 229 245 279 309 360 415 435 437 460 468 542   197
3.6 3.6 3.7 3.9 3.9 4.0 4.1 4.1 4.3 4.5 4.6 4.9 4.9 5.2 5.3   5.4 5.5 5.5 5.5 5.5 5.6 5.7 5.8 6.0 6.2 6.3 6.3 6.4 6.6   5.1

Atayal 12 20 28 37 75 81 91 106 107 120 130 141 196 201 250   253 257 269 277 278 282 297 303 306 320 328 364 508 777 .   221
3.8 4.0 4.1 4.1 4.6 4.7 4.8 4.9 4.9 5.0 5.1 5.1 5.4 5.4 5.6   5.6 5.6 5.7 5.7 5.7 5.7 5.8 5.8 5.8 5.8 5.9 6.0 6.5 8.2 .   5.4

Egypt 30 49 61 78 85 95 115 150 155 164 170 173 179 189 194   207 209 218 235 243 301 368 410 445 476 477 505 522 600 622   258
4.1 4.3 4.4 4.7 4.8 4.9 5.0 5.2 5.2 5.3 5.3 5.3 5.3 5.4 5.4   5.4 5.4 5.5 5.5 5.6 5.8 6.0 6.2 6.3 6.4 6.4 6.5 6.5 6.9 7.0   5.5

Berg 2 10 11 50 56 66 79 99 123 137 160 182 214 289 299   304 353 362 379 402 409 440 455 471 517 536 631 635 656 685   300
2.6 3.8 3.8 4.3 4.4 4.5 4.7 4.9 5.1 5.1 5.2 5.3 5.5 5.7 5.8   5.8 5.9 6.0 6.1 6.1 6.2 6.3 6.3 6.4 6.5 6.6 7.1 7.1 7.2 7.4   5.6

Teita 14 29 64 72 73 111 124 131 132 199 200 213 220 231 239   265 316 354 392 400 422 442 479 490 529 585 588 662 673 692   311
3.8 4.1 4.5 4.6 4.6 5.0 5.1 5.1 5.1 5.4 5.4 5.5 5.5 5.5 5.5   5.6 5.8 5.9 6.1 6.1 6.2 6.3 6.4 6.4 6.6 6.8 6.9 7.3 7.3 7.4   5.7

Philip 65 77 104 112 136 146 171 176 197 210 227 242 262 263 264   273 274 295 307 326 336 349 365 428 457 521 567 590 698 722   305
4.5 4.7 4.9 5.0 5.1 5.2 5.3 5.3 5.4 5.5 5.6 5.6 5.6 5.6 5.6   5.7 5.7 5.7 5.8 5.9 5.9 5.9 6.0 6.2 6.3 6.5 6.8 6.9 7.5 7.7   5.8

Peru 42 63 69 121 133 147 149 186 233 236 260 276 281 287 294   302 305 331 334 338 341 346 394 417 526 530 595 630 654 672   316
4.2 4.5 4.6 5.1 5.1 5.2 5.2 5.4 5.5 5.5 5.6 5.7 5.7 5.7 5.7   5.8 5.8 5.9 5.9 5.9 5.9 5.9 6.1 6.2 6.6 6.6 6.9 7.1 7.2 7.3   5.8

Austra 27 35 40 45 54 62 68 80 100 114 247 252 313 323 345   404 421 423 453 465 502 525 528 547 593 624 647 651 684 766   345
4.1 4.1 4.2 4.3 4.3 4.5 4.5 4.7 4.9 5.0 5.6 5.6 5.8 5.8 5.9   6.2 6.2 6.2 6.3 6.3 6.5 6.6 6.6 6.7 6.9 7.0 7.1 7.2 7.4 8.1   5.8

Tolai 25 47 57 74 87 97 126 128 151 193 240 248 268 285 292   310 324 355 377 380 384 401 485 573 608 644 693 705 739 792   336
4.0 4.3 4.4 4.6 4.8 4.9 5.1 5.1 5.2 5.4 5.6 5.6 5.7 5.7 5.7   5.8 5.9 5.9 6.1 6.1 6.1 6.1 6.4 6.8 7.0 7.1 7.4 7.5 7.8 8.4   5.9

Zulu 24 33 52 53 89 93 98 122 185 187 208 224 275 283 327   343 375 425 446 448 458 520 539 552 558 583 658 704 726 793   346
4.0 4.1 4.3 4.3 4.8 4.9 4.9 5.1 5.4 5.4 5.4 5.5 5.7 5.7 5.9   5.9 6.1 6.2 6.3 6.3 6.3 6.5 6.6 6.6 6.7 6.7 6.8 7.2 7.5 7.7   5.9

Ainu 4 76 90 119 138 142 178 198 203 204 308 312 319 342 347   383 396 399 433 447 456 532 535 556 562 572 574 580 661 671   358
3.3 4.7 4.8 5.0 5.1 5.2 5.3 5.4 5.4 5.4 5.8 5.8 5.8 5.8 5.9   6.1 6.1 6.1 6.2 6.3 6.3 6.6 6.6 6.6 6.7 6.7 6.8 6.8 7.2 7.3   5.9

Hainan 70 101 134 158 162 168 215 217 238 286 288 296 318 322 333   382 387 397 413 451 461 463 492 500 560 571 623 653 666 713   368
4.6 4.9 5.1 5.2 5.3 5.3 5.5 5.5 5.5 5.7 5.7 5.7 5.8 5.8 5.9   6.1 6.1 6.1 6.2 6.3 6.3 6.3 6.4 6.5 6.7 6.8 7.0 7.2 7.3 7.6   6.0

S.Cruz 43 55 103 139 143 157 172 184 225 230 246 255 293 314 351   366 454 470 474 486 510 527 531 582 601 616 675 736 741 759   381
4.2 4.4 4.9 5.1 5.2 5.2 5.3 5.4 5.5 5.5 5.6 5.6 5.7 5.8 5.9   6.0 6.3 6.4 6.4 6.4 6.5 6.6 6.6 6.8 6.9 7.0 7.3 7.8 7.9 8.0   6.1

Andama 127 135 165 192 195 205 219 254 259 270 329 344 352 369 386   403 418 480 504 518 537 568 578 581 591 669 706 708 709 752   417
5.1 5.1 5.3 5.4 5.4 5.4 5.5 5.6 5.6 5.7 5.9 5.9 5.9 6.0 6.1   6.1 6.2 6.4 6.5 6.5 6.6 6.8 6.8 6.8 6.9 7.3 7.5 7.5 7.5 7.9   6.2

Dogon 59 60 84 140 145 153 154 169 223 332 340 372 381 429 441   449 452 491 509 550 597 611 626 629 642 677 691 712 732 753   420
4.4 4.4 4.8 5.1 5.2 5.2 5.2 5.3 5.5 5.9 5.9 6.0 6.1 6.2 6.3   6.3 6.3 6.4 6.5 6.7 6.9 7.0 7.0 7.1 7.1 7.3 7.4 7.6 7.7 7.9   6.2

Bushma 58 82 125 241 256 258 272 284 300 311 317 325 337 339 373   395 472 516 541 545 555 587 602 605 632 634 646 659 727 740   424
4.4 4.7 5.1 5.6 5.6 5.6 5.7 5.7 5.8 5.8 5.8 5.9 5.9 5.9 6.0   6.1 6.4 6.5 6.6 6.6 6.6 6.7 6.8 7.0 7.0 7.1 7.1 7.1 7.2 7.7 7.9   6.2

Guam 38 113 148 159 163 232 266 267 335 357 371 412 416 427 431   439 450 467 481 499 503 544 596 604 618 621 627 676 762 765   426
4.2 5.0 5.2 5.2 5.3 5.5 5.5 5.7 5.9 6.0 6.0 6.2 6.2 6.2 6.2   6.3 6.3 6.4 6.4 6.5 6.5 6.6 6.9 7.0 7.0 7.0 7.1 7.3 8.0 8.1   6.3

Arikar 117 118 191 222 244 298 315 330 356 358 388 414 424 443 482   489 494 506 549 553 561 575 609 641 707 714 724 754 761 772   470
5.0 5.0 5.4 5.5 5.6 5.8 5.8 5.9 6.0 6.0 6.1 6.2 6.2 6.3 6.4   6.4 6.5 6.5 6.7 6.7 6.7 6.8 7.0 7.1 7.5 7.6 7.7 8.0 8.0 8.2   6.5

Tasman 86 94 129 166 211 261 291 348 359 376 378 405 419 464 475   501 511 514 565 619 620 636 640 643 667 718 720 735 750 828   471
4.8 4.9 5.1 5.3 5.5 5.6 5.7 5.9 6.0 6.1 6.1 6.2 6.2 6.3 6.4   6.5 6.5 6.5 6.8 7.0 7.0 7.1 7.1 7.1 7.3 7.6 7.7 7.8 7.9 9.3   6.5

N Japa 116 152 177 206 212 249 350 367 462 488 496 498 507 512 515   533 540 546 557 559 563 564 592 598 607 612 614 633 731 790   478
5.0 5.2 5.3 5.4 5.5 5.6 5.9 6.0 6.3 6.4 6.5 6.5 6.5 6.5 6.5   6.6 6.6 6.7 6.7 6.7 6.8 6.8 6.9 6.9 7.0 7.0 7.0 7.1 7.7 8.4   6.5

S Japa 48 202 234 271 370 390 391 407 411 444 459 478 483 484 497   513 538 551 554 584 599 650 680 696 719 734 744 745 789 807   516
4.3 5.4 5.5 5.7 6.0 6.1 6.1 6.2 6.2 6.3 6.3 6.4 6.4 6.4 6.5   6.5 6.6 6.7 6.8 6.9 7.2 7.3 7.5 7.6 7.7 7.9 7.9 8.4 8.7   6.7

Anyang 39 110 251 280 361 374 420 432 434 473 487 524 569 589 594   606 628 638 660 664 681 686 694 697 700 701 748 757 758 791   545
4.2 5.0 5.6 5.7 6.0 6.1 6.2 6.2 6.2 6.4 6.4 6.5 6.8 6.9 6.9   7.0 7.1 7.1 7.2 7.3 7.3 7.4 7.4 7.5 7.5 7.5 7.9 8.0 8.0 8.4   6.8

Easter 102 161 385 389 408 426 430 438 469 519 566 570 577 579 613   625 652 663 668 674 678 689 690 695 699 703 743 763 781 823   573
4.9 5.2 6.1 6.1 6.2 6.2 6.2 6.3 6.4 6.5 6.8 6.8 6.8 6.8 7.0   7.0 7.2 7.3 7.3 7.3 7.3 7.4 7.4 7.4 7.5 7.5 7.9 8.0 8.3 9.0   6.9

Morior 228 237 406 466 543 576 586 645 649 670 682 728 746 751 770   773 775 786 787 788 795 800 803 805 812 813 816 818 825 831   690
5.5 5.5 6.2 6.3 6.6 6.8 6.8 7.1 7.2 7.3 7.3 7.7 7.9 7.9 8.2   8.2 8.2 8.4 8.4 8.4 8.5 8.5 8.6 8.6 8.7 8.7 8.8 8.9 9.1 9.4   7.8

Eskimo 393 398 493 495 548 655 657 679 683 687 711 715 717 729 730   742 755 764 768 769 779 784 785 799 808 817 822 826 827 829   705
6.1 6.1 6.5 6.5 6.7 7.2 7.2 7.3 7.3 7.4 7.6 7.6 7.6 7.7 7.7   7.9 8.0 8.0 8.1 8.1 8.3 8.3 8.3 8.5 8.7 8.8 9.0 9.2 9.3 9.4   7.8

Buriat 174 290 436 534 603 610 665 702 710 733 737 749 767 778 782   794 802 806 814 815 820 821 824 830 832 834 835 836 837 839   720
5.3 5.7 6.3 6.6 7.0 7.0 7.3 7.5 7.6 7.7 7.8 7.9 8.1 8.3 8.3   8.4 8.6 8.7 8.7 8.8 8.9 9.0 9.1 9.4 9.5 9.7 9.8 9.8 10 11   8.3

Hawaii 615 617 637 639 648 688 716 721 723 725 738 747 756 760 771   774 776 780 783 796 797 798 801 804 809 810 811 819 833 838   751
7.0 7.0 7.1 7.1 7.2 7.4 7.6 7.7 7.7 7.7 7.8 7.9 8.0 8.0 8.2   8.2 8.2 8.3 8.3 8.5 8.5 8.5 8.6 8.6 8.7 8.7 8.7 8.9 9.6 10   8.1

pology. With these, the whole routine was followed to produce table 10, the POPKIN results. This makes Europeans, above all Zalavár, clearly the populations providing nearest neighbors, with generally acceptable distances for the means. However, Atayals are in the thick of it, if not as closely as Zalavár and Norse subjects. And

Filipinos, though providing few close neighbors, are overall not very distant, ranking about with the African Teita. Australoids are all further away and even more distant are Chinese and Japanese.

Here are the slightly different results produced by DISPOP:

Norse	Zalvar	Egypt	Atayal	Berg	Dogon	Ainu	Phil	Tasman	Teita	S. Cruz	Tolai
3.7	4.2	4.4	4.7	4.8	4.9	4.9	4.9	5.0	5.2	5.2	5.3
0.37	0.18	0.06	0.04	0.04	0.06	0.02	0.03	0.09	0.06	0.01	0.02

DISPOP thus makes the Norse sample, not Zalavár, the closest population, but with the same emphasis on Europe and modest possibility of affiliation for Atayals. The closest 11 populations, by distance, include no Japanese or Chinese but allow Tasmanians and Tolai as possibles.

This is a case, not of assigning an unknown modern skull to its ethnic niche, but one of getting a reading on a prehistoric individual whose context is important. Obviously we do not assign it to Europe, but rather ask what this means. In East Asia, nearness of Atayals is a comfortable reading as "proto-Mongoloid," or an older layer of such populations before expansion of Japanese and Han Chinese. The marked European affiliation has to point to a configuration that is more archaic for East Asians if not for Europeans. This is the point at which we may take note of the greater-than-Asian supraorbital development. Recall the earlier discussion of robusticity in Mesolithic North Africans.

Awatovi

This was a major Hopi town at the time of Spanish contact, but documentation was sketchy and made from a distance. It was more receptive to Christianity than were other Hopi pueblos, traditionally due to a miraculous cure of a native by a Spanish priest, and a sizable church was built in Awatovi. The Spanish were driven completely out of New Mexico in the Pueblo Revolt of 1680, and although Awatovi was revisited after the reconquest, there were no Spanish priests present between 1680 and 1700, when Awatovi was destroyed for good by hostile action

of uncertain origin. (Historical records were not made on the spot but in distant Santa Fe.) The ruins of the Franciscan church were excavated in the 1930s by the Peabody Museum, and burials found in it were brought back for study.

In 1989 Christy Turner recorded the dental characters of these remains, and left a note for me saying that one individual (PM catalogue N/3376) had a European dentition, suggesting that this was in fact a Spanish priest and asking if I would assess it metrically.

The specimen consists of a partial face and vault, lacking the occiput. Estimating a few of the variables by doubling the measurable half distance, it was possible to record a total of 33 measurements. As with Liujiang, these were used in a complete analysis from the beginning, ending with distances of the 839 pruned skulls from the Awatovi individual, producing the POPKIN results in table 11.

Here, the nearest neighbors are overwhelmingly Norse and Zalavár, which provide 23 of the 50 nearest individuals. American Indians are not the most distant among other populations overall, but provide very few close neighbors.

So, taking dental and metrical evidence together with the fact that the skull exhibits no signs of artificial deformation, otherwise almost universal at Awatovi, there can be little doubt that these are the remains of a Franciscan Spaniard. Only after reaching this point did I go to the publications on the excavation (Montgomery et al. 1949) to find to my chagrin that Turner and I had been anticipated by E. A. Hooton and the late Charles Snow. One burial from below the church altar was suspected, because of its circumstances, as being that of a Spaniard. Hooton

TABLE 11

Table 11

POPKIN results: Awatovi

```
Norse    1    2    3    5    6   10   11   12   13   15   18   25   29   37   41     55   56   61   65   77   90  129  133  152  160  186  407  412  420  565    107
        2.8  3.6  3.9  4.1  4.1  4.3  4.3  4.3  4.4  4.5  4.7  4.9  4.9  5.1  5.1    5.3  5.3  5.4  5.4  5.5  5.6  5.9  5.9  6.0  6.1  6.3  7.2  7.2  7.2  7.8    5.2

Zalava   4   14   16   17   23   38   39   47   53   54   57   60   85   87   88     98  113  119  135  138  146  159  181  187  225  279  308  361  389  410    131
        4.1  4.5  4.6  4.6  4.8  5.1  5.1  5.2  5.3  5.3  5.3  5.4  5.6  5.6  5.6    5.7  5.8  5.8  5.9  5.9  6.0  6.1  6.2  6.3  6.5  6.7  6.8  7.0  7.1  7.2    5.7

Egypt   28   42   62   69   75   79   81   82  101  103  106  109  111  115  193    211  215  228  240  275  301  313  331  342  352  473  474  536  625  679    232
        4.9  5.1  5.4  5.4  5.5  5.5  5.5  5.5  5.7  5.7  5.8  5.8  5.8  5.8  6.3    6.4  6.4  6.5  6.5  6.7  6.8  6.8  6.9  6.9  6.9  7.4  7.4  7.7  8.1  8.4    6.3

Peru    24   26   70   80   95  105  137  163  165  167  173  198  205  223  227    242  247  284  318  327  337  339  347  356  378  379  444  533  636  672    263
        4.8  4.9  5.4  5.5  5.7  5.7  5.9  6.1  6.1  6.2  6.2  6.3  6.3  6.5  6.5    6.6  6.6  6.7  6.8  6.9  6.9  6.9  6.9  7.0  7.0  7.0  7.3  7.7  8.2  8.4    6.5

Atayal   9   20   32   49   74   92  120  123  140  147  161  162  178  194  226    232  237  277  282  326  360  447  463  497  522  523  556  694  742     .    272
        4.3  4.8  5.0  5.3  5.6  5.8  5.8  5.9  6.0  6.1  6.1  6.2  6.3  6.5  6.5    6.5  6.5  6.7  6.7  6.9  7.0  7.3  7.4  7.5  7.6  7.6  8.5  9.0    .            6.5

Berg     8   21   40   43   46   58   59   78   94  114  196  202  219  234  249    288  291  305  325  359  402  409  422  430  433  449  499  632  644  671    274
        4.2  4.8  5.1  5.2  5.2  5.3  5.4  5.5  5.7  5.8  6.3  6.3  6.4  6.5  6.6    6.7  6.7  6.8  6.9  7.0  7.1  7.2  7.2  7.3  7.3  7.4  7.5  8.2  8.2  8.4    6.5

S Japa  22   30   50   72   89  104  116  130  134  155  200  203  216  222  250    264  270  278  299  328  340  351  362  398  453  468  529  628  670  772    280
        4.8  4.9  5.3  5.4  5.6  5.7  5.8  5.9  5.9  6.0  6.3  6.3  6.4  6.5  6.6    6.6  6.6  6.7  6.8  6.9  6.9  6.9  6.9  7.0  7.1  7.4  7.4  7.7  8.1  8.4  9.3    6.6

Anyang  33   36   73   96  112  127  150  184  188  214  243  257  260  261  268    300  330  363  401  405  406  426  487  491  506  531  588  602  662  740    325
        5.1  5.1  5.5  5.7  5.8  5.9  6.0  6.3  6.3  6.4  6.6  6.6  6.6  6.6  6.6    6.8  6.9  7.0  7.1  7.2  7.2  7.2  7.5  7.5  7.6  7.7  8.0  8.1  8.3  9.0    6.8

Hainan  34   71   76  126  136  151  175  192  204  218  238  241  252  304  319    320  355  366  373  393  413  425  454  459  525  527  530  548  631  713    326
        5.1  5.4  5.5  5.9  5.9  6.0  6.2  6.3  6.3  6.4  6.5  6.5  6.6  6.8  6.8    6.8  7.0  7.0  7.0  7.1  7.2  7.2  7.4  7.4  7.6  7.6  7.7  7.8  8.2  8.7    6.8

S.Cruz  45   52   68  122  142  143  157  229  236  255  266  280  297  315  335    354  357  369  375  376  394  404  416  428  508  561  571  578  621  655    330
        5.2  5.3 5.4  5.8  6.0  6.0  6.0  6.5  6.5  6.6  6.6  6.7  6.8  6.8  6.9    6.9  7.0  7.0  7.0  7.1  7.2  7.2  7.2  7.6  7.8  7.9  8.0  8.1  8.3    6.8

Ainu    31   91  107  110  139  156  168  220  221  231  248  263  274  287  298    317  323  344  380  387  438  440  460  469  493  547  563  624  626  651    330
        5.0  5.6  5.8  5.8  5.9  6.0  6.2  6.4  6.4  6.5  6.6  6.6  6.7  6.7  6.8    6.8  6.8  6.9  7.0  7.1  7.3  7.3  7.4  7.4  7.5  7.8  7.8  8.1  8.1  8.3    6.8

Arikar  19   84   93   97  108  128  158  179  201  209  233  283  296  303  365    388  396  397  429  443  450  472  479  492  513  537  584  592  622  726    343
        4.7  5.6  5.7  5.7  5.8  5.9  6.1  6.2  6.3  6.4  6.5  6.7  6.7  6.8  7.0    7.1  7.1  7.1  7.3  7.3  7.4  7.4  7.5  7.5  7.6  7.7  8.0  8.0  8.1  8.9    6.9

Guam    64   99  164  174  177  185  199  217  224  259  292  295  309  314  322    334  349  381  399  400  452  494  505  528  534  555  566  630  649  692    359
        5.4  5.7  6.1  6.2  6.2  6.3  6.3  6.4  6.5  6.6  6.7  6.7  6.8  6.8  6.8    6.9  6.9  7.0  7.1  7.1  7.4  7.5  7.6  7.6  7.7  7.8  7.8  8.2  8.3  8.5    7.0

Easter  86  100  124  169  230  235  271  272  290  312  332  336  343  350  372    403  411  414  457  471  478  483  518  541  545  586  683  698  710  725    398
        5.6  5.7  5.9  6.2  6.5  6.5  6.7  6.7  6.7  6.8  6.9  6.9  6.9  6.9  7.0    7.1  7.2  7.2  7.4  7.4  7.5  7.5  7.6  7.7  7.8  8.0  8.5  8.6  8.7  8.8    7.2

Morior  44   63  131  171  176  182  197  210  212  281  311  329  346  368  390    424  458  484  516  543  560  569  582  614  618  650  697  707  753  793    413
        5.2  5.4  5.9  6.2  6.2  6.3  6.4  6.4  6.4  6.7  6.8  6.9  6.9  6.9  7.0    7.2  7.4  7.5  7.6  7.8  7.8  7.8  7.9  8.0  8.1  8.1  8.3  8.5  8.7  9.1    7.2

Philip  48   51  118  141  154  195  206  262  338  374  385  395  423  439  455    467  470  476  500  503  504  551  573  583  587  609  659  686  696  746    420
        5.3  5.3  5.8  5.9  6.0  6.3  6.4  6.6  6.9  7.0  7.1  7.1  7.2  7.3  7.4    7.4  7.4  7.5  7.5  7.6  7.6  7.8  7.9  8.0  8.0  8.1  8.3  8.5  8.5  9.0    7.2

Teita  102  125  148  166  246  251  269  285  302  307  316  377  383  391  408    431  485  486  517  544  546  579  593  617  640  664  690  701  752  800    437
        5.7  5.9  6.0  6.1  6.6  6.6  6.6  6.7  6.8  6.8  6.8  7.0  7.0  7.1  7.2    7.3  7.5  7.5  7.6  7.8  7.8  8.0  8.0  8.1  8.2  8.3  8.5  8.6  9.1  9.7    7.4

Hawaii  27  117  121  213  253  289  294  310  324  345  421  434  441  462  475    511  514  524  549  553  580  604  605  634  677  714  718  731  791  809    468
        4.9  5.8  5.8  6.4  6.6  6.7  6.7  6.8  6.9  6.9  7.2  7.3  7.3  7.4  7.5    7.6  7.6  7.6  7.8  7.8  8.0  8.1  8.1  8.2  8.4  8.7  8.8  8.9  9.5  9.9    7.5

Eskimo   7  132  144  145  149  172  239  267  321  333  341  436  442  445  448    498  512  539  576  633  643  666  667  691  747  754  777  796  816  819    472
        4.2  5.9  6.0  6.0  6.0  6.2  6.5  6.6  6.8  6.9  6.9  7.3  7.3  7.3  7.4    7.5  7.6  7.7  7.9  8.2  8.2  8.3  8.3  8.5  9.0  9.1  9.4  9.6   10   10    7.6

N Japa  67  183  189  245  254  273  348  358  392  415  417  427  432  437  480    515  521  540  554  557  559  574  585  663  682  693  716  733  784  802    480
        5.4  6.3  6.3  6.6  6.6  6.6  6.9  7.0  7.1  7.2  7.2  7.2  7.3  7.3  7.5    7.6  7.6  7.7  7.8  7.8  7.9  8.0  8.3  8.5  8.5  8.7  8.9  9.5  9.8    7.6

Zulu   190  191  244  276  293  353  367  382  384  446  451  456  466  481  489    526  542  564  567  572  590  645  689  700  722  723  736  762  788  821    514
        6.3  6.3  6.6  6.7  6.7  6.9  6.9  7.0  7.0  7.1  7.3  7.4  7.4  7.4  7.5    7.5  7.6  7.7  7.8  7.8  7.9  8.0  8.3  8.5  8.6  8.8  8.8  8.9  9.2  9.5   10    7.8

Andama  35   83  153  306  490  509  538  552  568  570  589  591  596  598  608    616  619  627  635  641  647  648  687  695  715  735  748  757  764  785    570
        5.1  5.5  6.0  6.8  7.5  7.6  7.7  7.8  7.9  7.9  8.0  8.0  8.0  8.0  8.1    8.1  8.1  8.1  8.2  8.2  8.3  8.3  8.5  8.5  8.7  8.9  9.1  9.1  9.2  9.5    8.0

Tolai  256  258  364  370  418  465  488  495  562  607  610  612  637  639  642    646  660  680  699  706  712  728  739  749  755  758  765  778  805  810    614
        6.6  6.6  7.0  7.0  7.2  7.4  7.5  7.5  7.8  8.1  8.1  8.1  8.2  8.2  8.2    8.3  8.3  8.4  8.6  8.7  8.7  8.9  9.0  9.1  9.1  9.2  9.2  9.4  9.8   10    8.3

Bushma 170  180  207  265  461  507  550  581  594  615  658  669  673  675  681    688  703  708  727  730  734  741  750  760  774  780  782  786  799  836    626
        6.2  6.2  6.4  6.6  7.4  7.6  7.8  8.0  8.0  8.1  8.3  8.4  8.4  8.4  8.4    8.5  8.6  8.7  8.9  8.9  8.9  9.0  9.1  9.2  9.3  9.4  9.4  9.5  9.7   11    8.4

Buriat  66  286  371  419  435  501  520  532  600  620  638  674  676  685  709    711  721  732  745  767  769  770  776  804  808  814  826  831  833  835    649
        5.4  6.7  7.0  7.2  7.3  7.6  7.7  7.8  8.1  8.2  8.4  8.4  8.5  8.7    8.7  8.8  8.9  9.0  9.2  9.3  9.3  9.4  9.8  9.9   10   11   11   11   11    8.7

Dogon  477  496  502  519  558  595  597  599  601  613  623  629  652  653  717    729  738  744  759  768  781  792  794  811  813  815  822  823  830  838    690
        7.5  7.5  7.6  7.6  7.8  8.0  8.0  8.0  8.1  8.1  8.1  8.2  8.3  8.3  8.7    8.9  8.9  9.0  9.2  9.2  9.4  9.5  9.6   10   10   10   10   10   11   11    8.9

Austra 464  482  535  577  603  611  654  656  657  684  702  704  719  720  724    751  756  761  773  779  787  790  807  817  818  824  827  828  832  834    716
        7.4  7.5  7.7  8.0  8.1  8.1  8.3  8.3  8.3  8.5  8.6  8.6  8.8  8.8  8.8    9.1  9.1  9.2  9.3  9.4  9.5  9.5  9.9   10   10   10   11   11   11   11    9.1

Tasman 208  386  510  575  606  661  665  668  678  705  737  743  763  766  771    775  783  789  795  797  798  801  803  806  812  820  825  829  837  839    718
        6.4  7.1  7.6  7.9  8.1  8.3  8.3  8.4  8.4  8.7  8.9  9.0  9.2  9.2  9.3    9.4  9.5  9.5  9.6  9.6  9.7  9.8  9.8  9.9   10   10   10   11   11   12    9.2
```

examined the bones and his opinion supported this. Accordingly, the bones were sent to the Franciscan Fathers Mission to the Navajos in Arizona. The report continues, "At least one other skeleton of European physical type was encountered in the church. The anthropometric study of the bones is being made by Dr. Charles E. Snow and a detailed description of the burials will be published in the Awatovi Series of the Papers of the Peabody Museum." The report was not published, but Snow's records of observation are marked "Sp. priest" and the catalogue number, as well as measurements he took, identify the subject as the one here.

So we can only say that the objective approach of Turner and myself turns out to be as good as the eyeball assessment of Hooton and Snow. Actual identification of an individual from documents is not possible. One respected Franciscan, Father Porras, was martyred by poisoning, apparently courtesy of Awatovi native priests; his place of burial is not recorded but he was too old to have been either of the youthful individuals here described.

6 Other Workers, Other Work

Robustness of method, and some other final questions, can be inspected by comparison with approaches of other workers. This is mainly directed at some prehistoric circum-Mediterranean materials, dealt with by several of these workers, with the advantage of shedding more light on these populations themselves.

Present Study

DISPOP and POPKIN have produced these results:

Upper Paleolithic Europe: Mladeč 1 (Lautsch) is acceptable as a recent European, as possibly is Předmostí 4. Předmostí 3 and Chancelade are not.

Mesolithic: Afalou 9, of Algeria, seems acceptable as Zalavár (or Tasmanian or Berg); Afalou 5 is acceptable as modern, but not as European. The Teviec specimens from Brittany are acceptable as modern; Teviec 11 would be strongly Easter Island; Teviec 16 could be Zalavár or Ainu or Philippines.

Neanderthals, etc.: La Chapelle, La Ferrassie 1, and Shanidar 1 are rejected as anything modern, as is Skhūl 5.

CRANID

This is the interesting work and programming produced by Richard Wright (1992a, 1992b). The recent version (1992a) presents several procedures, using the same database (Howells 1989), selecting 29 measurements subjected to a principal components analysis and retaining components 2 to 21, the first being discarded as size-related.

One of Wright's procedures is IDCRAN. This is comparable to POPKIN but uses 2,524 individuals of the reference populations (here including Maoris), of both sexes, the individuals being ordered by distance from the target skull. The nearest 50 are listed and printed with their distances. (As a supplement, the whole space can be divided into segments, from 2 to 32 as desired, and the contents of the set nearest the target skull displayed.) The rationale is well set forth by Wright.

Quite possibly because of the use of principal components, I find on comparison that Wright's IDCRAN results are somewhat less focused than those from POPKIN. (I got similar effects when using principal components in studies herein.) For example, when the highly typical test Peruvian skull, *#1414*, is used as the target, then of the 50 nearest neighbors, my use of Wright's method gives 13 Peruvians versus 22 by POP-

KIN; and if all American skulls are counted, the nearest 50 include, by IDCRAN, 15, as against 42 by POPKIN. In general I find it difficult to read positive affiliation from IDCRAN; certainly it is difficult enough from POPKIN.

A second of Wright's procedures, MINI-CRAN, corresponds to DISPOP. This places a target skull among the means of all 55 samples (both sexes for 26 populations, plus the males of Anyang, Philippines, and the sample of Maori treated by me as test specimens.) From the results it creates dendrograms. Figure 4 shows the dendrogram of Wright's distances without a target skull included. This corresponds well to my own population dendrograms, e.g., figure 3 herein. Europeans, Americans, Africans, and Australoids group as might be expected, although Polynesians are distributed over other sets, with Easter Island anomalously appearing among Africans.

Figure 5, sections of such dendrograms, gives the positions of some target skulls as tested by me with MINICRAN. Of Neanderthals, La Chapelle appears as a somewhat distant element in the African-Australoid main branch. La Ferrassie 1, not shown, takes an identical position. Shanidar has the same relationship, but is less removed as a branch. Skhūl 5 has the same position and degree of removal.

For the Upper Paleolithic Předmostí 3 and 4 are both linked specifically with the Australoids. Mladeč 1, however, appears at the other end of the dendrogram, distantly linked to Europeans, and to Eskimos, who are here mathematically wrenched out of their position with Asiatics in figure 4. (This kind of disturbance is common in such clustering procedures.) It does not seem worthwhile to illustrate further cases but only to cite results obtained. Of the Mesolithic specimens, Afalou 9, like Mladeč 1, is distantly joined to Europeans, and in this case brings Buriats out of position; Afalou 5 specifically joins Tasmanians. Of the Brittany pair, Teviec 11, as in my results, affiliates with Easter Island (in its funny position among Africans) while Teviec 16 joins Moriori, Maori, and American Indians, i.e., moderns but not Europeans and not the other affiliates seen under POPKIN. MINICRAN does not make as great a distinction between "moderns" and earlier specimens as POPKIN tends to do.

In general, MINICRAN does not seem to be an improvement on findings herein, or on those in *Skull Shapes*. Figure 6, from figure 16 in the latter publication, presents a clustering, based on principal components from raw scores as in MINICRAN, and includes the main male populations plus a variety of other specimens or small samples. The Mesolithic Afalou and Teviec pairs and the Mesolithic Muge group of Portugal, here join Europeans at a slight remove. Some other prehistoric specimens are separated from the main body in two steps: one, which includes Mladeč 1 and Qafzeh 6, is distinct from the main body, while Neanderthals and Skhūl 5 are found in a group which is very much further removed from all.

The same result for Neanderthals appears in figure 7, taken from figure 15 in the previous work. Using Mahalanobis' D distances on the basic populations plus La Chapelle and La Ferrassie 1 as a sample of 2, this displays what might be looked on as a classic arrangement of groups which also sets Neanderthals well off from the body of modern populations.

As a matter of judgment, I feel that the methods herein, DISPOP and POPKIN, are somewhat more revealing that Wright's versions, and may in fact, from my whole experience, go about as far as is possible toward specific identification and relationships. In the future, perhaps better statistics and other information may go further, but the nature of the material, and its basic canons of variation, make it unlikely.

Van Vark et al.

G. N. van Vark, with associates, has for some years addressed metric comparisons of modern and late prehistoric crania, with attention to continuity and size changes, as have some others (e.g., Frayer 1984.) Again, the database for moderns is my own (1989). In a recent report (van Vark, Bilsborough, and Henke 1992) the authors use Mahalanobis' D^2, together with a precise statistic to determine the significance *between* two distances where this is useful. They assemble groups of prehistoric specimens as available, and as far as measurements in incomplete cases allow grouping. The approach is not, as here, to deter-

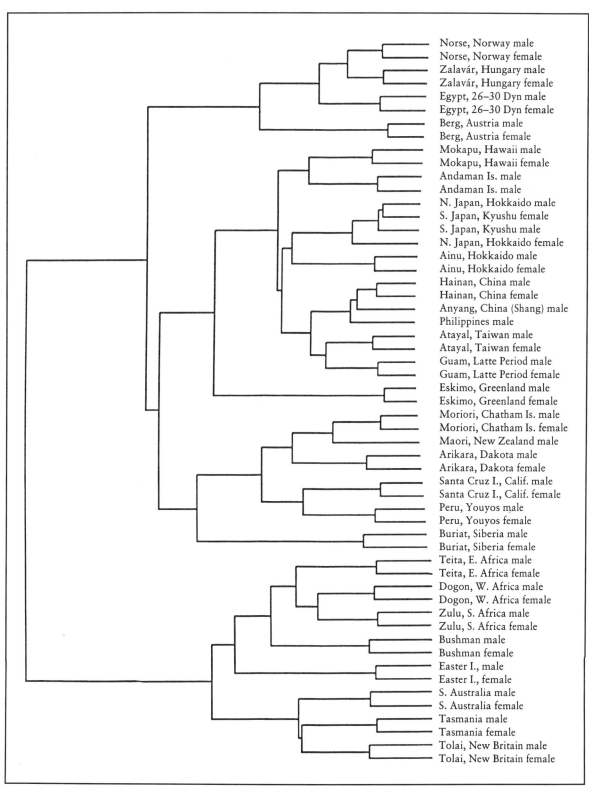

Figure 4. Dendrogram of 55 samples, male and female, generated by Wright's MINICRAN.

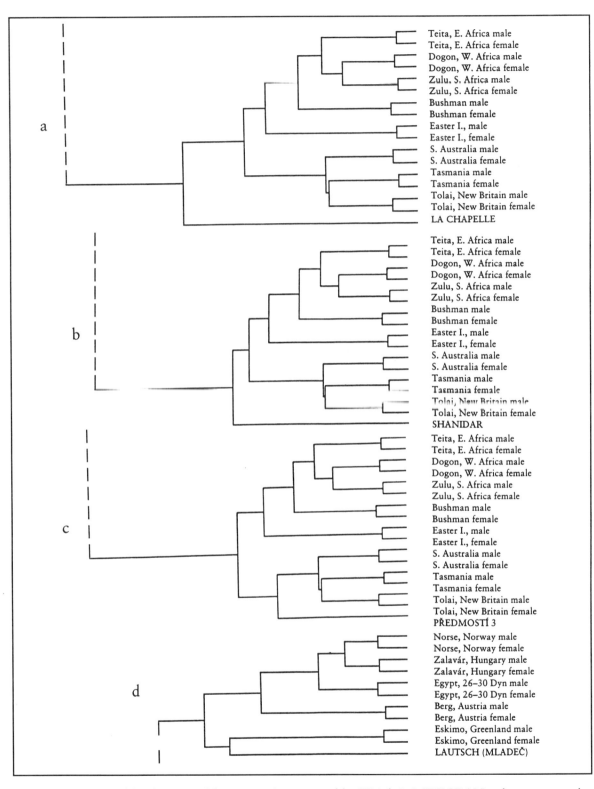

Figure 5. Sections of dendrograms like Figure 4, generated by Wright's MINICRAN, when computed to include a) La Chapelle, b) Shanidar 1, c) Předmostí 3, and d) Mladeč 1.

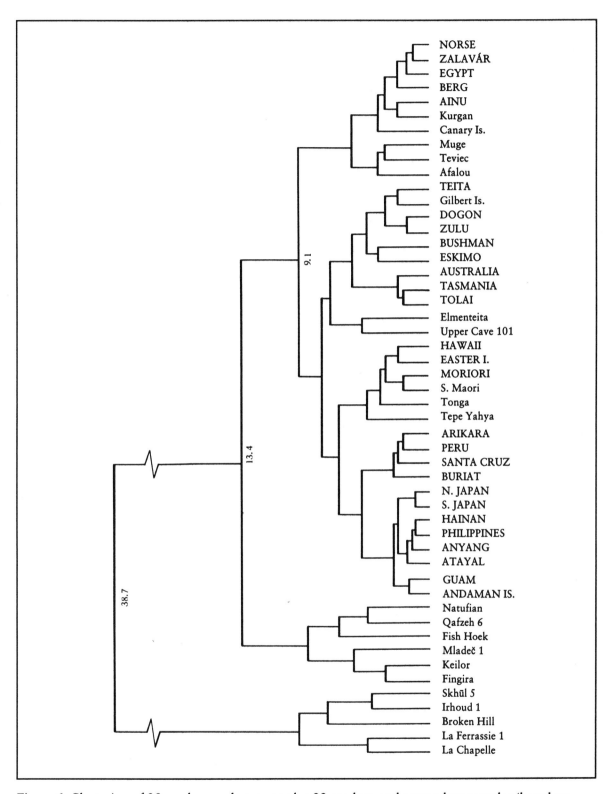

Figure 6. Clustering of 28 modern male groups plus 22 modern and premodern samples (based on principal components analysis of raw scores).

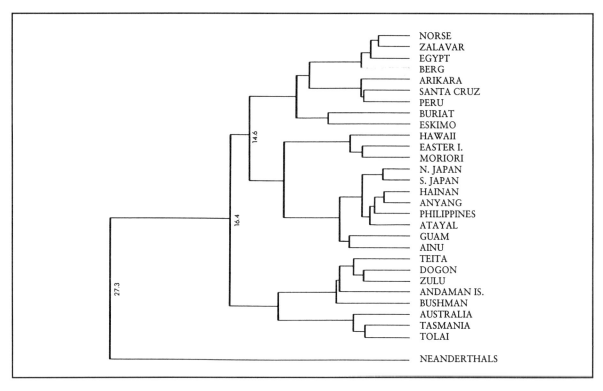

Figure 7. Clustering of 28 male groups plus 2 Neanderthals (based on Mahalanobis D distances).

mine affiliation of target skulls, but results bear strongly on problems faced herein. Such results are the following.

The general recent sample differed from the Mesolithic (Teviec, Mugem) and the Upper Paleolithic (European), with the Paleolithic sample being significantly more distant from the modern than was the Mesolithic. In one test, when the modern sample was limited to Europeans, this test of differences was not significant.

When the Upper Paleolithic sample was divided into early and late, the early was significantly more different from the moderns, though not in all tests.

Samples of Asiatic and African *Homo erectus* were decidedly more distant from moderns than was the Upper Paleolithic sample.

In this study the sample (7) of European Neanderthals was also decidedly more distant from moderns than was the Upper Paleolithic sample, whether the moderns were represented by the total sample or by Europeans only. Separate dis-

tances were computed for measurements of the face and of the braincase. For the latter, Neanderthals were slightly less distant from Europeans, early or late, than from other regional groups (especially Africans). For the face, Europeans early or late were more distant from the Neanderthals than were other modern regions. Only American Indians were equally as distant as Europeans in the tests using face measures.

It should be emphasized that this is a study of distances and thus of differences. It concerns mainly the steps of difference in Europe from moderns back to the earliest Upper Paleolithic, in what suggests to the authors a "gradual anagenetic change" (as contrasting with what they see as an abrupt disappearance of Neanderthals). But it is not suggested from the figures, or by the authors, that the European Paleolithic group is closer to modern Europeans than to other populations; in fact, it may have closer morphometric affinity to some peripheral groups (no figures given), as do the Neanderthals as well.

Kidder et al.

Kidder, Jantz, and Smith (1992) look at the same problem, testing whether a number of individual crania, essentially Late Pleistocene European, can be included in modern *Homo sapiens*. They do this by creating a reference sample to represent moderns, composed of our 1989 Arikara, Peru, Tolai, and Zalavár series (both sexes) plus series of "American Whites" and "American Blacks." By log transformations, they arrive at size and shape measures (up to 17 measurements are available) which are used to provide D^2 distances from the reference modern sample. Distances within a 95 percent confidence limit are noted as allowable "moderns." Several analyses were performed, according to sets of measurements available because of damage or incompleteness in fossils.

All Neanderthals are especially distant *except* for the early Saccopastore 1, which is within modern limits. Skhūl 4, Skhūl 5, and Qafzeh 6 are outside; Qafzeh 9 falls within.

Unlike the DISPOP or POPKIN results, 3 Mladeč specimens fall outside while 5 Předmostí specimens fall within and 4 Cro Magnons divide. Later Paleolithic crania largely fall within. Thus there are signals of change toward moderns, with earlier specimens representing stages still outside moderns, as defined by multivariate representation of a large modern sample.

C. L. Brace

C. Loring Brace (1991) has made a limited investigation of some of these connections, using his own set of 24 measurements emphasizing the circumnasal skeleton. He used discriminant analysis covering ten modern populations or groupings measured by himself, to estimate the probability that a given fossil could be a member of each of them. He reported that, by the percentage probabilities, the Monte Circeo Neanderthal could be excluded from all modern series except the European, and likewise that Qafzeh 6 could be allowed only as African (or, with small probability, Australian).

These conclusions are at variance with other results reported herein. It should be remarked that his percentages are evidently posterior prob-

abilities. That is, the set of all probabilities sums to 1.0, so that the least distant case may have a high apparent "probability" when the actual distance is considerable and the true probability remote. Thus Brace's suggestions are not persuasive. I have discussed this earlier, on page 3. Albrecht (1992) and van Vark have pointed out the limitations of the procedure.

Such disagreements remind us that methods and input play a large part in outcomes, and that convincing results will come only with time and with fully appropriate methods. For example, though not of appropriateness, I have run a simple discriminant analysis to discriminate "Europeans" from "Non-Europeans," i.e., the three European samples plus Egypt against the lumped remaining series. In the results, all my Neanderthals were "Non-European" as were Skhūl 5, Irhoud, Afalou 5, and, less positively, the Teviec pair. Mladeč 1, Předmostí 3, Afalou 9, and Chancelade were "European," all with posterior probabilities of .99 to 1.00. I do not consider this weighty evidence. See below.

Regional Classification

DISPOP and POPKIN have used specific and localized (ideally) populations as reference samples. As a practical aim we might want to identify a target skull more generally as "European" versus "African" versus "American" or, more baldly, by "race" as implied by geography. (See Giles and Elliott 1962, for a first and much-used attempt. Also see Brues 1992, on the problems of such approaches). To that end I have used the procedures of DISPOP to assign crania to one of six main regions, by pooling appropriate populations to form regional reference samples. In the results, it is apparent that this is broadly effective but has shortcomings that comment forcefully on the hazards of unwary reliance on single or inappropriate solutions.

The regional samples were made up by these combinations of the male populations:

Europe: Norse, Zalavár, Berg
Africa: Teita, Dogon, Zulu
Australo-Melanesia: Australia, Tasmania,Tolai
Polynesia: Hawaii, Easter Island, Moriori
Asia: South Japan, Hainan, Anyang
America: Arikara, Santa Cruz, Peru

Six groups allow only 5 canonical variates. Use of these to classify the 894 members of the reference samples themselves resulted, *using posterior probabilities*, in a 97.5 percent correct classification to region, a result not surprising when applied to the subjects used to derive the discriminating variates originally. See remarks on page 2, relative to a paper by Hershkovitz et al.

When the male TEST specimens were used, results agreed very generally with those shown under DISPOP: assignment of Melanesians, Polynesians, and Africans to the groups expected, and including the affiliation of the Peruvians #1403 and #1435 as "Australo-Melanesian" and "European" respectively. It is not worth exploring all this here, except for two kinds of departures from DISPOP results, exemplified in table 12. As displayed, the closest five populations found by DISPOP (from tables 4 and 5) are on the left, with "regional" distances to the right.

The first kind of difference from DISPOP involves recent peoples. Four Avars of Hungary are European, Asian, or American, without the connection to Buriats, highly suggestive if not close, seen under DISPOP. Three Chukchis persuasively become "Asian" or "European" rather than the more appropriate "Eskimo" of DISPOP. Madrasis, Veddas, and Burmans appear as Africa or Europe (one American), without the suggestion of other possibilities, notably the Andamans connection for Veddas suggested by DISPOP. In an egregious failure, various Jomon skulls of Japan, generally affiliated with Ainu under DIS-

POP, are "European" or "American" here (results not shown). By contrast, the numerous specimens from Yayoi burials, representing the founding population of later Japan, are all "Asia."

Two things are suggested by these and similar cases. First, when a possibly appropriate population is not involved (e.g., Buriats, Andamanese, Ainus) assignment is necessarily defective and deceptively simple. Second, a regional assignment tends to generalize over the populations appearing closest in DISPOP; not necessarily a bad result, but, as in the first case, less discriminating than is ideally possible.

The second kind of departure from DISPOP may be allied to the above but involves prehistoric specimens. As above, Fish Hoek, firmly Bushman in other tests, is here, with no Bush in the reference framework, either European or Asian, not African. So the difficulty of placing the Elmenteita, Afalou, and Teviec specimens, seen earlier and repeated here, comes to the fore again: robusticity? or lack of kin among the reference populations? I consider either to be plausible. As to Upper Cave 101, no fresh light is shed but the darkness is not deepened.

But older specimens, more deviant in shape from moderns, give anomalous results. Qafzeh 6 is acceptably Australo-Melanesian, and not African as Brace finds, showing again that analogous tests can give quite different answers. But here the identification is less surprising than the computed probability of that identification.

TABLE 12

Regional assignments compared to DISPOP results

	DISPOP					*Regional*					
1565 Avar	Buriat	Berg	Eskimo	Guam	Zalvar	Europe	Asia	America	Austral	Polynes	Africa
	4.1	7.6	7.7	8.0	8.0	2.7	4.6	4.7	6.3	6.6	6.7
	0.25	–	–	–	–	0.20	0.00	–	–	–	–
1566 Avar	Buriat	Teita	Norse	Anyang	Hainan	Asia	Europe	Africa	America	Polynes	Austral
	5.3	5.7	5.8	6.1	6.1	3.0	3.4	4.0	4.8	6.4	6.6
	0.01	0.00	0.00	0.00	0.00	0.09	0.03	0.00	–	–	–
1567 Avar	Buriat	Eskimo	S.Cruz	N.Jap.	Zalvar	America	Asia	Europe	Austral	Polynes	Africa
	5.2	6.2	6.3	6.6	6.6	1.8	3.1	3.5	4.4	4.6	5.2
	0.02	0.00	0.00	–	–	0.66	0.07	0.02	0.00	0.00	–
1568 Avar	S.Cruz	Zalvar	Arikar	Norse	Peru	America	Europe	Austral	Asia	Africa	Polynes
	4.0	4.5	4.5	4.6	4.7	2.8	3.5	3.6	3.7	4.6	5.5
	0.29	0.12	0.11	0.08	0.07	0.16	0.02	0.02	0.01	0.00	–

TABLE 12 CONTINUED

Regional assignments compared to DISPOP results

1667 Chukchi

DISPOP

Eskimo	Guam	S.Jap.	N.Jap.	Arikar
3.7	4.9	5.2	5.3	5.3
0.48	0.04	0.01	0.01	0.01

Regional

Asia	Polynes	America	Europe	Africa	Austral
2.4	3.2	3.4	4.3	5.7	5.7
0.35	0.06	0.03	0.00	-	-

1668 Chukchi

DISPOP

Eskimo	Norse	Zalvar	Guam	Arikar
4.9	5.3	5.4	5.5	5.6
0.04	0.01	0.01	0.01	0.00

Regional

Europe	America	Asia	Africa	Polynes	Austral
2.5	3.4	3.6	3.8	3.9	4.2
0.29	0.03	0.02	0.01	0.00	0.00

1669 Chukchi

DISPOP

Eskimo	Zalvar	Guam	Norse	Arikar
4.9	6.3	6.5	6.5	6.7
0.05	0.00	-	-	-

Regional

Europe	America	Asia	Austral	Africa	Polynes
2.1	3.2	3.8	4.0	4.4	4.9
0.52	0.06	0.01	0.00	0.00	-

1615 Madras

DISPOP

Teita	Egypt	Zalvar	Norse	Andamn
4.5	4.7	5.0	5.5	5.6
0.11	0.08	0.03	0.00	0.00

Regional

Europe	Africa	Asia	Austral	America	Polynes
3.4	3.6	4.6	5.2	5.7	6.9
0.03	0.01	0.00	-	-	-

1616 Madras

DISPOP

Zalvar	Egypt	Norse	Berg	Phil
5.0	5.2	5.3	5.5	5.6
0.03	0.02	0.01	0.01	0.00

Regional

Europe	Africa	Asia	Polynes	America	Austral
2.6	4.2	4.6	5.6	5.7	5.9
0.26	0.00	0.00	-	-	-

1619 Vedda

DISPOP

Andamn	Phil	Egypt	Zalvar	Atayal
3.5	4.6	4.8	4.9	5.0
0.57	0.09	0.06	0.04	0.03

Regional

Europe	Africa	Austral	Asia	America	Polynes
3.4	3.7	4.0	4.2	4.7	6.4
0.03	0.01	0.00	0.00	0.00	-

1620 Vedda

DISPOP

Egypt	Andamn	Dogon	Teita	Phil
4.1	4.1	4.9	4.9	5.3
0.25	0.24	0.04	0.04	0.01

Regional

Africa	Asia	Europe	Austral	America	Polynes
2.6	3.5	3.7	3.8	4.5	5.0
0.24	0.02	0.01	0.01	0.00	-

1623 Burma

DISPOP

Peru	Arikar	Phil	Zalvar	Hainan
5.5	5.5	5.6	5.8	6.0
0.01	0.01	0.00	0.00	0.00

Regional

America	Europe	Asia	Austral	Africa	Polynes
2.1	3.5	4.1	4.9	5.5	6.2
0.53	0.02	0.00	-	-	-

1624 Burma

DISPOP

Egypt	Teita	Zalvar	Norse	Zulu
4.3	4.4	4.7	4.9	4.9
0.17	0.16	0.07	0.05	0.05

Regional

Africa	Austral	Europe	America	Asia	Polynes
2.4	3.4	3.9	4.6	5.0	5.1
0.35	0.03	0.00	0.00	-	-

1625 Burma

DISPOP

Phil	Andamn	Guam	Egypt	Hainan
5.0	5.2	5.4	5.7	5.8
0.03	0.02	0.01	0.00	0.00

Regional

Europe	America	Asia	Polynes	Africa	Austral
2.7	2.8	3.3	4.2	4.3	5.3
0.18	0.17	0.05	0.00	0.00	-

1626 Burma

DISPOP

Egypt	Zulu	Teita	Zalvar	Norse
3.9	5.1	5.2	5.3	5.5
0.33	0.02	0.02	0.01	0.00

Regional

Africa	Europe	Asia	America	Austral	Polynes
3.3	3.4	5.3	6.1	6.7	7.0
0.04	0.04	-	-	-	-

1703 Fish Hoek

DISPOP

Bush	Norse	Zalvar	S.Cruz	Berg
3.6	6.8	6.8	7.0	7.1
0.50	-	-	-	-

Regional

Europe	Asia	Africa	America	Austral	Polynes
3.0	3.8	3.9	4.6	5.3	6.7
0.11	0.01	0.00	0.00	-	-

2718 Upper Cave 101 Cast

DISPOP

Arikar	Zalvar	Easter	Norse	Berg
6.0	6.1	6.2	6.2	6.2
0.00	0.00	0.00	0.00	0.00

Regional

Europe	Polynes	America	Austral	Asia	Africa
3.7	3.8	4.7	4.9	6.1	6.7
0.01	0.01	-	-	-	-

2083 Djebel Quafzeh 6 Orig.

DISPOP

Tasman	Austr	Tolai	Ainu	Zulu
4.7	5.5	5.5	5.7	6.2
0.08	0.01	0.01	0.00	0.00

Regional

Austral	Polynes	Africa	America	Europe	Asia
2.6	4.0	4.9	5.2	6.0	6.0
0.22	0.00	-	-	-	-

2080 Djebel Irhoud 1 Cast

DISPOP

Tasman	Austr	Tolai	S.Cruz	Arikar
8.1	9.2	9.8	10.3	10.3
-	-	-	-	-

Regional

Austral	Africa	America	Europe	Polynes	Asia
2.0	5.4	5.6	6.8	6.8	7.3
0.57	-	-	-	-	-

2082 La Chapelle Orig.

DISPOP

Tasman	Austr	Morior	S.Cruz	Tolai
9.1	10.3	10.8	10.9	10.9
-	-	-	-	-

Regional

Austral	America	Polynes	Africa	Europe	Asia
1.5	4.7	5.2	5.6	5.9	6.5
0.83	-	-	-	-	-

2903 La Ferrassie Cast

DISPOP

Tasman	Arikar	Morior	S.Cruz	Berg
7.0	7.2	7.3	7.4	7.6
-	-	-	-	-

Regional

America	Austral	Europe	Polynes	Africa	Asia
2.8	3.7	4.3	5.1	5.2	5.9
0.17	0.01	0.00	-	-	-

1778 Shanidar Cast

DISPOP

Arikar	Berg	Tasman	Morior	Zalvar
7.7	8.1	8.2	8.3	8.6
-	-	-	-	-

Regional

America	Europe	Austral	Polynes	Asia	Africa
2.5	2.9	4.3	4.5	5.3	5.9
0.31	0.12	0.00	0.00	-	-

1712 Broken Hill 1 Orig.

DISPOP

Austr	Tasman	S.Cruz	Tolai	Arikar
8.3	8.3	9.1	9.6	9.7
-	-	-	-	-

Regional

Austral	America	Europe	Africa	Asia	Polynes
2.3	3.9	5.5	5.7	6.5	6.5
0.38	0.01	-	-	-	-

Under DISPOP the remaining cases in table 12 have no probability at all of assignment to a modern population. But here, Irhoud 1, La Chapelle, and Broken Hill have ridiculously high probabilities of being "Australo-Melanesian" while La Ferrassie 1 and Shanidar can readily pass as "American"! The readings are simply not acceptable (compare figs. 5, 6, and 7), and these anomalous findings may be given a little consideration.

The reasons are probably technical, and due to the smaller number of canonical variates used (5 versus 14 in DISPOP), with the special form of these specimens not engaged by these lowest order variates. The two analyses are on some-what different data sets, of course, but results are similar over most of the test specimens. A detailed examination of the computation of scores and distances for the La Chapelle skull (not shown here) seems to show a) large C-score deviations, plus or minus, as would be expected from the low vault and prominent midface of this Neanderthal, and b) the cumulative scores on 5 canonical variates are in each case relatively minor deviations from zero. The conclusion: in the context of discrimination based on generally similar modern populations, *without including higher order variates*, such radically different shapes can, and do, give almost random and meaningless answers. If this is right, it is one more comment on the limits of modern cranial data as a context for earlier specimens.

CLUSTERING THE INDIVIDUALS

In general, the populations of this study, as constituted, act like natural units, although not totally discrete ones, with the overlapping we have seen. On the other hand the attempt above, to construct regional or "racial" groups or units, like "Caucasoid" by pooling Europeans, have not been successful, being too rigid to encompass the much broader variation that we clearly observe. Both these things are probably reflecting, incompletely, the actual nature of the distribution of the variation in cranial form among moderns.

Another view of the same things may be seen by referring to the section called "Sorting the Individuals" in my 1989 report. This exhibits the FASTCLUS treatment of the material. The individuals, from their C-scores, are allowed to form clusters without regard to their population of origin. (The approach is analogous to that under POPKIN.) In these results, and others not published, the individuals were sorted so as to bring together the populations of a given region while not confining the individuals of a population strongly to a single such cluster. (The most isolated, in this sense, tended to be Buriats and Eskimos—lacking regional brothers or sisters?—but also Berg, Easter Island, and Australians.)

The 1989 tabulations sorted all the individuals, and allowed the number of clusters formed to be of the order of 4 to 7. Table 13 provides another such clustering, here of the 839 males of the samples of 30 as used in POPKIN, and based on their C-scores. The number of possible clusters is set at 28.

This suggests both the tendency to segregate and to overlap. Europeans, with Egypt, tend to confine themselves to 3 clusters, mainly one dominated by Berg and one by the other three. In departures, a few Norse and Zalavár join a cluster (#23) clearly that of Ainus (who do not reciprocate). In Africa, Bushmen are isolated, while Dogon tend to cluster separately from Bantu-speaking Teita and Zulu. Six East Asian peoples are found mainly in an overlapping block of 4 clusters, while Guam is distributed all over the place. By contrast, Andamanese, Buriats, and Eskimos are well isolated, though the first and the last each generate two clusters. As a special note, the relative isolation of the Bush and Tasmanians may support the legitimacy of the selection of the individuals that compose them, which was something of a problem originally (see Howells 1973).

These details are not important, and another selection of samples would probably match this only broadly. The significant thing is the balance of clustering (dominant) and overlapping (considerable).

TABLE 13

A clustering treatment of the "pruned" populations

	Cluster																											
	1	2	3	4	5	6	7	8	9	10	11	12	13	14	15	16	17	18	19	20	21	22	23	24	25	26	27	28
Norse	21	3	1			1											1						3					
Zalavar	12	8	4			1																	4		1			
Berg	2	9	19																									
Egypt	26		1			1																		1	1			
Teita	1			14	9			5		1																		
Dogon				5	1	16			1								1			1		2		1	3			
Zulu				16	1	3				1										5		1	1		2			
Bushman				3	1		25		1																			
Australia				1				23	3	3																		
Tasmania								2	26	2																		
Tolai								9	2	17				1				1										
Hawaii											17	12	1															
Easter I												5	25															
Moriori											1			20	9													
Arikara														3	1	12	11	1	2									
Santa Cruz		1				2				1								3	22	1		1	1					
Peru	3		2			2													17	6								
N Japan			1								1	4		2				1	13	5	2	1						
S Japan			1	1	2							3		2					2	9	8	2						
Hainan			.		2	1												3		10	2	5	4		2	1		
Anyang												5						1		5	1	10	8					
Atayal	1		3		1													2		8	1	4	1					
Philippines					1				1		1							1		3	2	15	1	1	4			
Guam					1						1	4	2	2	4	3			3	1	3	1	1		1	1		3
Ainu									1		1			1	1						2	24						
Andaman						2												1					15	12				
Buriat			1													1										28		
Eskimo														1													18	11
CLUSTER N	66	21	33	40	17	29	25	39	34	25	21	34	28	27	20	16	24	42	50	28	34	45	36	20	24	29	18	14

7 General Summary

Altogether the results, as addressed to the simple purpose of estimating affiliation of a modern skull, are very good. This is apparent, for example, in assignment of known Peruvian tests skulls, as well as suggestion of affiliation as "Polynesian" for Maori cases, or as "Melanesian" (Tolai) for many specimens from central Melanesia, or with finding all TEST Eskimos to be in fact "Eskimo," and so on. In addition, negative evidence can be strong. One might come into court and say "No, I am 95 percent certain this is not the skull of Hiawatha."

Among moderns there is clearly a small percentage of flatly wrong assignments. Similarly, there is a lack of useful indication for specimens from areas not represented among the base series, e.g., India and southern Indonesia, and probably much of the Americas.

Beyond actual recent peoples matters change somewhat. Relatively late prehistoric specimens confirm expectable affiliations in many cases; in others the assignment is unreasonable. Certain earlier cases, like Mladeč 1, seem to fall into place among modern populations of an area. However, such specimens as Afalou 5, Teviec 11, Elmenteita A and B, and Upper Cave 101 all are generally recognized as modern anatomically but are here probabilistically well removed, while suggesting affiliations which are not credible. They seem to represent the common situation.

In the preceding section, my own results as well as those of others are seen as suggesting that, typically, these late prehistoric specimens represent a stage just outside modern humanity cranially, for whatever reason, such as robusticity. This is perhaps not the only reason, but it points again to the mutual cranial closeness of modern populations generally.

As to the earlier parts of the Pleistocene, specimens from the fringes of *Homo sapiens sapiens* to Neanderthals and other archaics are seemingly too far removed from the multivariate common space of recent populations to give readable assignments. Typically, they lie at distances of greater than 7 standard deviations from any recent population at all. Evidently, attempts to affiliate archaic *Homo*, like Neanderthals, with modern populations are futile using the data and methods herein.

That is, all this work has little to say about premodern phylogeny. This is where studies by van Vark, using a different structure and methods, are more informative. Chris Stringer (see Stringer 1994) has returned to these problems over two decades, assembling samples and measurements as available for earlier *Homo*. His careful work is instructive for phylogenetic questions in ways not possible for the material and analyses of this study.

Finally, it is my opinion that the methods herein are at least as good for the purposes stated as are other methods suggested, and that for the present have been carried about as far as possible. They result from a good deal of trial and experiment. Possible combinations and modes of attack are of course infinite.

BEARINGS ON ORIGINS

This has not been a systematic study of prehistoric materials, but only an opportunistic coverage of a number of available specimens, and one requiring a full set of measurements on each. Work by Stringer and van Vark, just mentioned, has explored incomplete specimens to much greater effect. Nevertheless a number of implications as to the origin of moderns can be drawn here.

In the first place, the general mutual closeness of all modern populations, apparent in *Skull Shapes*, is confirmed herein. (This includes Australians, who, in the work of M. Lahr, were somewhat isolated.) The very fact of clear errors of identification of modern test specimens, made across continents, supports this general homogeneity.

Recent known or probable history is reflected at points. The westward pressure of "Mongoloid" peoples can be noted in the earlier presence of Europoids in the eastern Siberian Kurgan burials and the Mongoloid aspect of Avar skulls from Hungary. Late prehistoric skulls in Kenyan sites like Elmenteita appear as distinctly less "Negroid" than present Bantu-speakers of the same locality.

Certain crania of the latest Pleistocene exhibit connections with the present in some areas: Europe, Southern Africa, Australia (Keilor). In other regions, specimens of equivalent age fail to show such connections, notably Zhoukoudian Upper Cave 101.

This seems to mark a kind of boundary. There are few signs in the above evidence of continuities with anything earlier anywhere—certainly not with specimens approaching Neanderthals in archaic morphology. The evidence appears simply to uphold modern homogeneity only a short distance into the past. There are some cases of plausible modernity, like Mladeč 1 or Liujiang or Qafzeh 9 (the last not tested by me.)

From all this and what follows, I think that Afalou and Teviec are not necessarily to be rejected as "Caucasoid" forerunners, nor even Upper Cave 101 as having such a relation to "Mongoloids." This embattled skull clearly does not affiliate with any of our fairly numerous Asian populations, but may be declaring, not that it is European or American or Polynesian but simply that, like the more recent Afalou or Teviec persons, it is out of the reach of a completely modern conformation.

CRANIAL FORM: "RACE," REGION, AND POPULATION STRUCTURE

As above, identification of test specimens with specific populations has been quite good. There is, however, plenty of interdigitation, as when a certified Peruvian, #1435, refuses to be recognized as anything but European. Also, dendrograms of these 28 populations repeatedly group them well, as units, by geography, or by expected "racial" membership. However there is a certain fluidity. Andamanese move back and forth between Africa and Pacific peoples, and Polynesians may affiliate inconsistently (see Howells 1989.)

As all the foregoing shows, the populations as constituted act like natural units, although not totally discrete ones. More broadly, what can we infer regarding the texture of the craniometric variation in all the data? A generation ago, in a note entitled "On the non-existence of human races," Frank Livingstone (1962) pointed out that human variability "does not conform to the discrete packages named races." He epitomized his point in an often quoted aphorism: "There are no races, there are only clines." This drew an accompanying rejoinder from Th. Dobzhansky to the effect that he was smoothing things out too much.

I would say: "There are no races, there are only populations." As detailed above I have had mixed success in distinguishing *regions* craniometrically, that is, in setting up regional samples to which single individuals can be assigned as successfully as they can be affiliated with specific samples. Nonetheless such samples cluster themselves well in accordance with regional expectation. The populations are the genetic units. They can be grouped satisfactorily by region in dendrograms, but on the other hand the individuals assign themselves to specific populations better than to "races" or regional samples. And when there is no local population (Southeast Asia) to receive them, individual allocations seem to be erratic.

In a way this is simply blood groups revisited. In the early days of application of blood genetics to "race" (e.g., Boyd 1950), attempts to define major blood group "races" were not persuasive. Only recently, as Cavalli-Sforza (1991) has used more specific populations in cluster analyses, have affiliations like those from craniometry been apparent.

THE SCALE OF CRANIOMETRIC VARIATION

All this seems to accord with a recency of specific cranial form among populations, and a general community of moderns. Indeed, from what is known archaeologically, it would appear that certain population cranial differences among Polynesians are of decidedly recent origin and if data were available, American populations would probably provide a broader spectrum of craniometric variation.

This kind of thing is susceptible to better arithmetic statement. The cranial data of this study have been used by Relethford (1994) in more complex analyses to register variation between populations in both craniometric and genetic traits and in a further study (Relethford and Harpending 1994) to take account of the effects of population size and migration on amonggroup variation with their implications for recent human origins. Relethford proceeds by estimates of the ratio of among-group variation to total species variation. The estimates are roughly equal for craniometric and genetic data, and low in both, leading further to the surprising conclusion of negligible selective influence on either overall. (By contrast, skin color as measured by reflectance scores provides a much higher estimate of relative among-group variation, pointing to the expected strong effects of natural selection.)

These revealing studies again suggest a general closeness among modern populations. Of course, this can be overdone. There is clearly a certain cohesion within existing or recent regional peoples. There is no evident cranial overlap, for example, between Polynesians and Melanesians, an important matter. But the general picture of local population variation is one that was not apparent, or at least not accepted, in the early days of craniology.

Addendum

In this treatise I have made suggestions, at several points, that cranial robustness is present as a special factor confusing possible affiliations of certain prehistoric crania (e.g., Afalou). The suggestion was simply a surmise, without objective support other than the study by Marta Lahr (1992) in which she found a general distinction of her Afalou/Taforalt group by greater size and robusticity (see page 43). Now she and Richard Wright, extending her 1992 study, have produced a thorough and inventive analysis, addressed specifically to the relations of robusticity with size and shape, by using both measurements and non-metric ratings of external features reflecting robustness (especially facial), and employing multivariate expressions like canonical correlations and principal components. I cannot do justice here to this detailed and expressive work.

Among many things they discern a general complex of robusticity (of related, rather than independent, external features) largely determined by absolute measured size (though in Australians also partly by shape, i.e., a long narrow skull). For example, their few Polynesian and Ainu specimens appear as robust in the determi-

nations. These two populations are in fact particularly large-skulled, as can be seen from mean figures in *Skull Shapes*; however, in my work here, direct evidence of size is vitiated by the use of C-scores, and so robusticity can only be inferred, as I have done, without direct translation from one study to the other.

The authors deduce that generalized robustness was more characteristic of late Pleistocene, but anatomically modern, *Homo sapiens*, and that in recent times such a level of robusticity has receded in all populations (the Australians being special), so that robusticity is without particular phylogenetic significance. Specifically, they find that the Afalou-Taforalt specimens of North Africa, and those of the Upper Cave at Zhoukoudian, are in the large/robust region of variation (with Liujiang centered among the 10 Afalou/Taforalt specimens! although no phylogenetic connections are implied). Not included in their study are Teviec or the East African Elmenteita crania, where the influence of size or robustness was suggested by me, supra, as causing perturbations in the relations of such crania that DISPOP might otherwise have been expected to reveal.

Bibliography

Albrecht, G.
1992 "Assessing the Affinities of Fossils Using Canonical Variates and Generalized Distances." *Human Evolution* 7:49–69.

Anthony, D. W.
1985 "The 'Kurgan Culture,' Indo-European Origins and the Domestication of the Horse: a Reconsideration." *Current Anthropology* 27(4):291–313.

Boulinier, G.
1970- "L'Ordinateur Va-t-il Permettre un Renou-
1971 veau de l'Anthropologie?" *Bulletin de la Société Suisse d'Anthropologie et d'Ethnologie* 47:23–42.

Brace, C. L.
n.d. "Modern Human Origins and the Dynamics of Regional Continuity." Paper prepared for Prehistoric Mongoloid Dispersals Symposium, University of Tokyo, 1992.
1991 "Monte Circeo, Neanderthals, and Continuity in European Cranial Morphology: a Rear End View," in *Il cranio Neandertaliano Circeo 1. Studi e documenti*, 175–195. Museo Nazionale Preistorico Etnografico "Luigi Pigorini," Rome.

Brace, C. L., and K. D. Hunt
1990 "A Non-racial Craniofacial Perspective on Human Variation: A(ustralia) to Z(uni)." *American Journal of Physical Anthropology* 82:341–360.

Brace, C. L., and D. P. Tracer
1992 "Craniofacial Continuity and Change: a Comparison of Late Pleistocene and Recent Europe and Asia," in *The Evolution and Dispersal of Modern Humans in Asia*, T. Akazawa, K. Aoki, and T. Kimura, eds., 439–475. Hokusen-sha, Tokyo.

Brues, A. M.
1992 "Forensic Diagnosis of Race: General Race vs. Specific Populations." *Social Science & Medicine* 34:125–128.

Bulbeck, F. D.
1981 "Continuities in Southeast Asian Evolution Since the Late Pleistocene." Master's thesis. Canberra.

Cavalli-Sforza, L. L.
1991 "Genes, Peoples and Languages." *Scientific American* 265(5):104–110.

Chai, C. K.
1967 *Taiwan Aborigines: A Genetic Study of Tribal Variations*. Harvard University Press, Cambridge.

105

Clark, J. D., and S. A. Brandt (eds.)
1984   *From Hunters to Farmers: The Causes and Consequences of Food Production in Africa.* University of California Press, Berkeley.

Coon, C. S.
1939   *The Races of Europe.* Macmillan, New York.
1962   *The Origin of Races.* Alfred A. Knopf, New York.

da Cunha, E.
1989   "Cálculo de Funções Discriminantes para o Diagnose Sexual do Crânio." Ph.D. dissertation. Institute of Anthropology, University of Coimbra, Coimbra.

da Cunha, E., and G. N. van Vark
1991   "The Construction of Sex Discriminant Functions from a Large Collection of Known Sex." *International Journal of Anthropology* 6:53–66.

Ferencz, M.
1991   "Comparisons of Avar Anthropological Series." *Human Evolution* 6(2):213–223.

Frayer, D. W.
1984   "Biological and Cultural Change in the European Late Pleistocene and Early Holocene," in *The Origins of Modern Humans: A World Survey of the Fossil Evidence*, F. H. Smith and F. Spencer, eds., 211–250. Alan R. Liss, New York.

Giles, E., and O. Elliot
1962   "Race Identification from Cranial Measurements." *Journal of Forensic Sciences* 7(2):147–157.
1963   "Sex Determination by Discriminant Function Analysis of Crania." *American Journal of Physical Anthropology* 21(1):53–68.

Hedges, R. E. M., R. A. Housley, C. Bronk-Ramsey, and G. J. Van Klinken
1992   "Radiocarbon Dates from the Oxford AMS system: Archaeometry Datelist 14." *Archaeometry* 34: 155.

Hershkovitz, I., B. Ring, and E. Kobyliansky
1990   "Efficiency of Cranial Bilateral Measurements in Separating Human Populations." *American Journal of Physical Anthropology* 83:307–319.

Hooton, E. A.
1925   *The Ancient Inhabitants of the Canary Islands.* Harvard African Studies, vol. VII. Peabody Museum, Harvard University, Cambridge.

Howells, W. W.
1937   *Anthropometry of the Natives of Arnhem Land and the Australian Race Problem. Based on Data of W. L. Warner.* Papers of the Peabody Museum of Archaeology and Ethnology, vol. XVI, no. 1. Peabody Museum, Harvard University, Cambridge.
1966   "The Jomon Population of Japan. A Study by Discriminant Analysis of Japanese and Ainu Crania," in *Craniometry and Multivariate Analysis*, 1-43. Papers of the Peabody Museum of Archaeology and Ethnology, vol. LVII, no. 1. Peabody Museum, Harvard University, Cambridge.
1970   "Anthropometric Grouping Analysis of Pacific Peoples." *Archaeology & Physical Anthropology in Oceania* 5(3):192–217.
1973   *Cranial Variation in Man: A Study by Multivariate Analysis of Patterns of Difference Among Recent Human Populations*, Papers of the Peabody Museum of Archaeology and Ethnology, vol. 67. Peabody Museum, Harvard University, Cambridge.
1983   "Origins of the Chinese People: Interpretations of the Recent Evidence," in *The Origins of Chinese Civilization*, D. N. Keightley, ed., 297-319. University of California Press, Berkeley.
1986   "Physical Anthropology of the Prehistoric Japanese," in *Windows on the Japanese Past*, Richard J. Pearson, ed., 85-89. Michigan Center for Japanese Studies, Ann Arbor.
1989   *Skull Shapes and the Map: Craniometric Analyses in the Dispersion of Modern Homo.* Papers of the Peabody Museum of Archaeology and Ethnology, vol. 79. Peabody Museum, Harvard University, Cambridge.
n.d.   "Oceania," in *History of Physical Anthropology: An Encyclopedia*, F. Spencer, ed.

Kamminga, J., and R. V. S. Wright
1988   "The Upper Cave at Zhoukoudian and the Origin of the Mongoloids." *Journal of Human Evolution* 17:739–767.

Kidder, J. H., R. J. Jantz, and F. H. Smith
1992   "Defining Modern Humans," in *Continuity or Replacement. Controversies in Homo sapiens Evolution*, G. Bräuer and F. H. Smith, eds., 157–177. Balkema, Rotterdam.

Klein, R. G.
1989   *The Human Career: Human Biological and Cultural Origins*. University of Chicago Press, Chicago.

Knußman, R.
1967   "Penrose-Abstand und Diskriminanzanalyse," *Homo* 18(3):134–140.

Kritscher. H., and J. Szilvássy
1988-   "Populationsanalyse einer frügeschichtlichen
1989   Bevölkerung am Beispiel des awarischen Gräberfeldes von Zwölfaxing, NÖ." *Mitteilungen der Anthropologischen Gesellschaft in Wien* 118/119:345–368.

Lahr, M. M.
1992   "The Origins of Modern Humans: A Test of the Multiregional Hypothesis." Ph.D. dissertation. University of Cambridge.

Lahr, M. M., and R. V. S. Wright
n.d.   "The Question of Robusticity and the Relationship Between Cranial Size and Shape." *Journal of Human Evolution*. In press.

Leakey, L. S. B.
1935   *The Stone Age Races of Kenya*. Oxford University Press, London.

Livingstone, F.
1962   "On the Non-existence of Human Races." *Current Anthropology* 3:279-281

Montgomery, R. G., W. Smith, and J. O. Brew
1949   *Franciscan Awatovi. The Excavation and Conjectural Reconstruction of a 17th Century Spanish Mission Establishment at a Hopi Indian Town in Northeastern Arizona*. Papers of the Peabody Museum, vol. XXXVI. Peabody Museum, Harvard University, Cambridge.

Mukherjee, R., C. R. Rao, and J. C. Trevor
1955   *The Ancient Inhabitants of Jebel Moya (Sudan)*. Occasional Papers of the Cambridge Museum of Archaeology and Ethnology, vol. III, xi, 123. Cambridge.

Neumann, G. K.
1959   "Race, Language and Culture in Aboriginal North America." Abstract of unpublished paper presented at 28th Annual Meeting of American Association of Physical Anthropologists. *American Journal of Physical Anthropology* 18:362.

Ossenberg, N. S.
1994   "Origins and Affinities of the Native Peoples of Northwestern North America: the Evidence of Cranial Nonmetric Traits," in *Method and Theory for Investigating the Peopling of the Americas*, eds. R. Bonnichsen and D. Gentry Steele. Peopling of the Americas Publications. Edited Volume Series. Center for the Study of the First Americans, Oregon State University, Corvallis, OR.

Pardoe, C.
1991   "Isolation and Evolution in Tasmania." *Current Anthropology* 12(1):1–31.

Rao, C. R.
1948   "The Utilization of Multiple Measurements in Problems of Biological Classification." *Journal of the Royal Statistical Society* Series B, 10:159–203.

Relethford, J. H.
1994   "Craniometric Variation among Modern Human Populations." *American Journal of Physical Anthropology* 95: 53–62.

Relethford, J. H., and H. C. Harpending
1994   "Craniometric Variation, Genetic Theory, and Modern Human Origins." *American Journal of Physical Anthropology* 95: 249–270.

Rightmire, G. P.
1984   "Human Skeletal Remains from Eastern Africa," in *From Hunters to Farmers: The Causes and Consequences of Food Production in Africa*, J. D. Clark and S. A. Brandt, eds., 191–199. University of California Press, Berkeley.

Smith, F. H.
1984   "Fossil Hominids from the Upper Pleistocene of Central Europe and the Origin of Modern Europeans," in *The Origins of Modern Humans: A World Survey of the Fossil Evidence*, F. H. Smith and F. Spencer, eds., 137–209. Alan R. Liss, New York.

Stringer, C. B.
1994 "Out of Africa—A Personal History," in *Origins of Anatomically Modern Humans*, Matthew H. Nitecki and D. V. Nitecki, eds., 149–172. Plenum Press, New York.

van Vark, G. N.
1984 "On the Determination of Hominid Affinities," in *Multivariate Statistical Methods in Physical Anthropology*, G. N. van Vark and W. W. Howells, eds., 323–249. D. Reidel, Dordrecht.

van Vark, G. N., and J. Dijkema
1988 "Some Notes on the Origin of the Chinese People." *Homo* 39:143–148.

van Vark, G. N., A. Bilsborough, and W. Henke
1992 "Affinities of European Upper Palaeolithic *Homo sapiens* and Later Human Evolution." *Journal of Human Evolution* 23(5):401–417.

Weidenreich, F.
1939 "On the Earliest Representatives of Modern Mankind Recovered on the Soil of East Asia." *Peking Natural History Bulletin* 13:161–180.

Wolpoff, M. H.
1994 Review of *The Evolution and Dispersal of Modern Humans* (by T. Akazawa et al., eds.) *American Anthropologist* 96:184–187.

Wolpoff, M. H., Wu X. -z., and A. G. Thorne
1984 "Modern *Homo sapiens* Origins: a General Theory of Hominid Evolution Involving the Fossil Evidence from East Asia," in *The Origins of Modern Humans: A World Survey of the Fossil Evidence*, F. H. Smith and F. Spencer, eds., 411–483. Alan R. Liss, New York.

Wright, R. V. S.
1992a "Identifying the Origin of a Human Cranium: Computerised Assistance by CRANID." Sydney.
1992b "Correlation between Cranial Form and Geography in Homo sapiens: CRANID — A Computer Program for Forensic and Other Applications." *Archaeology in Oceania* 27(3):128–134.